SpringerBriefs in Statistics

JSS Research Series in Statistics

Editors-in-chief

Naoto Kunitomo
Akimichi Takemura

Series editors

Genshiro Kitagawa
Manabu Iwasaki
Tomoyuki Higuchi
Nakahiro Yoshida
Yutaka Kano
Toshimitsu Hamasaki
Shigeyuki Matsui

The current research of statistics in Japan has expanded in several directions in line with recent trends in academic activities in the area of statistics and statistical sciences over the globe. The core of these research activities in statistics in Japan has been the Japan Statistical Society (JSS). This society, the oldest and largest academic organization for statistics in Japan, was founded in 1931 by a handful of pioneer statisticians and economists and now has a history of about 80 years. Many distinguished scholars have been members, including the influential statistician Hirotugu Akaike, who was a past president of JSS, and the notable mathematician Kiyosi Itô, who was an earlier member of the Institute of Statistical Mathematics (ISM), which has been a closely related organization since the establishment of ISM. The society has two academic journals: the Journal of the Japan Statistical Society (English Series) and the Journal of the Japan Statistical Society (Japanese Series). The membership of JSS consists of researchers, teachers, and professional statisticians in many different fields including mathematics, statistics, engineering, medical sciences, government statistics, economics, business, psychology, education, and many other natural, biological, and social sciences.

The JSS Series of Statistics aims to publish recent results of current research activities in the areas of statistics and statistical sciences in Japan that otherwise would not be available in English; they are complementary to the two JSS academic journals, both English and Japanese. Because the scope of a research paper in academic journals inevitably has become narrowly focused and condensed in recent years, this series is intended to fill the gap between academic research activities and the form of a single academic paper.

The series will be of great interest to a wide audience of researchers, teachers, professional statisticians, and graduate students in many countries who are interested in statistics and statistical sciences, in statistical theory, and in various areas of statistical applications.

More information about this series at http://www.springer.com/series/13497

Toshio Sakata · Toshio Sumi
Mitsuhiro Miyazaki

Algebraic and Computational Aspects of Real Tensor Ranks

Springer

Toshio Sakata
Faculty of Design
Kyushu University
Fukuoka
Japan

Mitsuhiro Miyazaki
Department of Mathematics
Kyoto University of Education
Kyoto
Japan

Toshio Sumi
Faculty of Arts and Science
Kyushu University
Fukuoka
Japan

ISSN 2191-544X   ISSN 2191-5458   (electronic)
SpringerBriefs in Statistics
ISSN 2364-0057   ISSN 2364-0065   (electronic)
JSS Research Series in Statistics
ISBN 978-4-431-55458-5   ISBN 978-4-431-55459-2   (eBook)
DOI 10.1007/978-4-431-55459-2

Library of Congress Control Number: 2016934008

Printed on acid-free paper

This Springer imprint is published by Springer Nature
The registered company is Springer Japan KK

Preface

Recently, multi-way data or tensor data have been employed in various applied fields (see Mørup 2011). Multi-way data are familiar to statisticians as contingency tables. The difference between tensor data and contingency tables is that a contingency table describes count data, and therefore, its entries are necessarily integers, whereas the entries of a tensor datum are real numbers. The main feature of data analysis is often to decompose a datum into simple parts and extract its main parts. For example, Fourier analysis of a signal decomposes the signal into many parts with different frequencies and extracts the main frequencies contained in the signal. Similarly, we consider the decomposition of a tensor datum into a sum of rank-1 tensors, where rank-1 tensors are considered to be the simplest tensors. The minimal length of the rank-1 tensors in the sum is called the rank of the tensor. The objective of rank determination is to answer the question, "How many rank-1 tensors are required to express the given tensor?" In other words, we must find the simplest structure in a given datum. Thus, tensor rank is important for data analysts. In matrix theory, rank plays a key role and expresses the complexity of a matrix. Similarly, tensor rank is considered as an index of the complexity of a tensor. However, it is difficult to determine the tensor rank of a given tensor even for tensors of small size. Tensor rank also depends on the basis field; for example, the rank may be different in the cases of the complex number field $\mathbb{C}$ and the real number field $\mathbb{R}$. Many researchers have explored tensor ranks over the complex number field, where the property of algebraic closedness of the complex number field often makes the theory clear or easy.

In this book, we focus on the rank over the real number field $\mathbb{R}$, which is particularly interesting for statisticians. Rank-1 decomposition was first introduced by Hitchcock 1927, and he referred to it as a polyadic form. Subsequently, several authors investigated tensor rank, including Kruskal (1977), Ja'Ja' (1979), Atkinson and Stephens (1979), Atkinson and Lloyd (1980), Strassen (1983), and ten Berge (2000). In recent years, interest in tensor rank has been rekindled among several mathematicians, including Kolda and Bader (2009), de Silva and Lim (2008), Friedland (2012), Landsberg (2012), De Lathauwer et al. (2000), and Ottaviani

(2013). In addition, nonnegative tensors have been the subject of many studies on applied data analysis (see, for example, Cichocki et al. 2009). In this expository book, we mainly treat maximal rank and typical rank of real 2-tensors and real 3-tensors, and we summarize our research results obtained over nearly eight years since 2008. The maximal rank of size (m, n, p) tensors is the largest rank of tensors of this size, whereas the typical rank of size (m, n, p) tensors is a rank such that the set of tensors of rank r has a positive measure. Here, we re-emphasize that this book treats both tensor rank and typical ranks over the real number field.

This book is organized into eight chapters. Chapter 1 presents the terminologies and basic notions. Chapter 2 introduces propositions that characterize tensor rank. In consideration of beginners or novices, Chap. 3 treats simple and ad hoc evaluation methods of tensor rank by column and row operations as well as matrix diagonalization for tensors of small size ($2 \times 2 \times 2$, $2 \times 2 \times 3$, $2 \times 3 \times 3$, and $3 \times 3 \times 3$). Chapter 4 introduces an absolutely nonsingular tensor and a determinant polynomial. In addition, it discusses the relation between (i) the existence of absolutely nonsingular tensors and Hurwitz-Radon numbers and (ii) absolutely full column tensors and bilinear forms. Chapter 5 treats the maximal rank of $m \times n \times 2$ and $m \times n \times 3$ tensors. Chapter 6 treats generic ranks and typical ranks of quasi-tall tensors. Chapter 7 presents an overview of the global theory of tensor rank and discusses the Jacobian method. Finally, Chap. 8 treats $2 \times 2 \times \cdots \times 2$ tensors.

Toshio Sakata
Toshio Sumi
Mitsuhiro Miyazaki

Contents

Chapter 1
Basics of Tensor Rank

In this chapter we introduce the basic concepts of tensor rank.

1.1 Tensor in Statistics

In statistical data analysis, a tensor is a multi-way array datum. Just as the complexity of a matrix datum is described by its matrix rank, the complexity of a tensor datum is described by its tensor rank. In this chapter, we review several fundamental concepts of tensor rank. Typical rank and maximal rank are treated in later chapters. A historical reference is Hitchcock (1927). Useful introductory references for the basics of tensor rank include Kolda and Bader (2009), De Lathauwer et al. (2000), and Lim (2014). For further reading, refer to Ja'Ja' (1979), Kruskal (1977), Strassen (1983), ten Berge (2000), Comon et al. (2009), and Landsberg (2012). For tensor algebra, we refer to the book by Northcott (2008). First, we define a tensor datum over a basis field $\mathbb{F}$.

Definition 1.1 A multi-way array $T = (T_{i_1 i_2 \ldots i_K})$, $1 \leq i_1 \leq N_1, \ldots, 1 \leq i_K \leq N_K$, is called a K-way tensor with size $(N_1, N_2, \ldots, N_K)$.

Remark 1.1 For short, we use "$N_1 \times \cdots \times N_K$ tensor" instead of "a tensor T of size $(N_1, \ldots, N_K)$", especially when K is small.

Definition 1.2 The set of K-way tensors with size $(N_1, N_2, \ldots, N_K)$ over $\mathbb{F}$ is denoted by $T_{\mathbb{F}}(N_1, \ldots, N_K)$ or simply $\mathbb{F}^{N_1 \times \cdots \times N_K}$.

In this book, we consider the case of $\mathbb{F} = \mathbb{C}$ and $\mathbb{R}$, and we omit $\mathbb{F}$ from the suffixes when there is no scope for confusion. Hence, the set $T_{\mathbb{F}}(N_1, \ldots, N_K)$ is often denoted as $T(N_1, \ldots, N_K)$ without confusion. Further, note that $\mathbb{F}^{N_1 \times \cdots \times N_K}$ is equal to $\mathbb{F}^{N_1 \ldots N_K}$ as a set.

For statisticians, a tensor as a multi-way array is familiar as a higher-order contingency table. The difference is that a contingency table takes integer values as elements, whereas an array tensor takes arbitrary real values, complex values, or elements of an arbitrary field $\mathbb{K}$.

© The Author(s) 2016
T. Sakata et al., *Algebraic and Computational Aspects of Real Tensor Ranks*,
JSS Research Series in Statistics, DOI 10.1007/978-4-431-55459-2_1

A tensor is a multi-array datum (this is a 3-tensor)

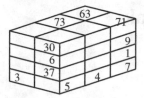

On the other hand, for mathematicians, a tensor is familiar as an element of a tensor product of vector spaces.

Definition 1.3 Let $V_i = \mathbb{F}^{N_i}$ with a fixed basis $\{v_{i1}, \ldots, v_{iN_i}\}$, where $v_{ij} = (0, \ldots, 0, 1, \ldots, 0)^T$ (1 in the jth position) for $1 \leq i \leq K$. Then, the tensor product $V_1 \otimes V_2 \otimes \cdots \otimes V_K$ is a vector space over $\mathbb{F}$ with a basis $\{v_{1i_1} \otimes v_{2i_2} \otimes \cdots \otimes v_{Ki_k} | 1 \leq i_j \leq N_j, j = 1, 2, \ldots, K\}$ and the elements of the tensor products $V_1 \otimes V_2 \otimes \cdots \otimes V_K$ are called K-mode tensors.

The two concepts of multi-way tensors in statistics and tensors in algebra are mutually exchangeable. Since $\{v_{1i_1} \otimes v_{2i_2} \otimes \cdots \otimes v_{Ki_k} | 1 \leq i_j \leq N_j, j = 1, 2, \ldots, K\}$ is a basis of $V_1 \otimes V_2 \otimes \cdots \otimes V_K$ over $\mathbb{F}$, there is a one-to-one correspondence between $T_{\mathbb{F}}(N_1, \ldots, N_K)$ and $V_1 \otimes V_2 \otimes \cdots \otimes V_K$, where $T = (T_{i_1 i_2 \ldots i_K})$ corresponds to $\sum_{i_1=1}^{N_1} \cdots \sum_{i_K=1}^{N_K} T_{i_1 \ldots i_K} v_{1i_1} \otimes \cdots \otimes v_{Ki_K}$. For example, a tensor $(a_1, a_2) \otimes (b_1, b_2) \otimes (c_1, c_2)$ corresponds to a $2 \times 2 \times 2$ array tensor $T = (T_{ijk}) = (a_i b_j c_j)$. Under this identification, we treat an element of $V_1 \otimes V_2 \otimes \cdots \otimes V_K$ and an $N_1 \times \cdots \times N_K$ multi-way tensor reversibly and call both of them simply as a K-tensor without confusion. Note that we are concerend mainly with 3-tensors.

Now, we define rank-1 tensors.

Definition 1.4 A nonzero K-tensor $T = (T_{i_1 \ldots i_K})$ is called a rank-1 tensor if $T = (T_{i_1 \ldots i_K}) = (a_{i_1 1} \ldots a_{i_K K})$ for some vectors $a_1, \ldots, a_K$.

Below, we illustrate a 3-tensor of rank 1.

A rank-1 tensor

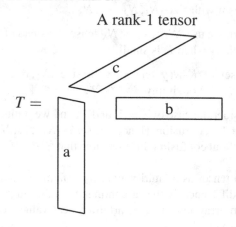

Here, we note that any K-tensor $T = (T_{i_1 \ldots i_K})$ can be expressed as a set of $(K - 1)$ tensors as follows:

$$T = (T_1; \ldots; T_K), T_k$$
$$= (T_{i_1 \ldots i_{K-1}k} | 1 \leq i_1 \leq N_1, \ldots, 1 \leq i_{K-1} \leq N_{k-1}), \quad 1 \leq k \leq K.$$

In particular, when $K = 3$, any 3-tensor $T = (T_{i_1 \ldots i_K})$ can be expressed as slices of matrices:

$$T = (T_1; \ldots; T_K), T_k = (T_{i_1 i_2 k} | 1 \leq i_1 \leq N_1, \text{ and } 1 \leq i_2 \leq N_2), \quad 1 \leq k \leq K.$$

This is called a slice representation along the third axis of a 3-tensor T and is often used in subsequent sections. For a 3-tensor, a slice representation along the first and second axes are also defined similarly.

Example 1.1 A rank-1 tensor T with size $2 \times 2 \times 2$ is a tensor $T = (T_{i_1 i_2 i_3})$, where $T_{i_1 i_2 i_3} = a_{i_1} b_{i_2} c_{i_3}$, $1 \leq i_1 \leq 2$, $1 \leq i_2 \leq 2$, $1 \leq i_3 \leq 2$ for some two-dimensional vector $a = (a_1, a_2)$, $b = (b_1, b_2)$, $c = (c_1, c_2)$. For example, when $a = (1, 2)$, $b = (3, 4)$, $c = (5, 6)$,

$$T = a \otimes b \otimes c = \left(\begin{pmatrix} 15 & 20 \\ 30 & 40 \end{pmatrix} ; \begin{pmatrix} 18 & 24 \\ 36 & 48 \end{pmatrix} \right)$$

is a rank-1 tensor.

Example 1.2 If a size (N_1, N_2, N_3) tensor $P = (p_{ijk})$ expresses a joint probability function of three discrete random variables, the independence model is equivalent to P being a rank-1 tensor.

The PARAFAC model for a tensor decomposes the tensor into a sum of rank-1 tensors.

Definition 1.5 For a tensor $T = (T_{i_1 \ldots i_K}, 1 \leq i_1 \leq N_1, \ldots, 1 \leq i_K \leq N_K)$, the PARAFAC model describes (decomposes) T as $T = T_1 + \cdots + T_s$, where T_i is a rank-1 tensor.

Tensor rank is defined as follows.

Definition 1.6 For a K tensor $v \in V = V_1 \otimes \cdots \otimes V_K$, the minimum integer r such that there is an expression $v = v_1 + \cdots + v_r$, where $v_i \in V$ are rank-1 tensors, is called the tensor rank of v and is denoted by $\mathrm{rank}_{\mathbb{F}}(v)$. Correspondingly, for a K-tensor T, the minimum integer r such that there is an expression $T = T_1 + \cdots + T_r$ where T_i are rank-1 tensors, is called a tensor rank of T and denoted by $\mathrm{rank}_{\mathbb{F}}(T)$.

Below, we illustrate a rank-r 3-tensor.

Tensor decomposition into a sum of rank-1 tensors

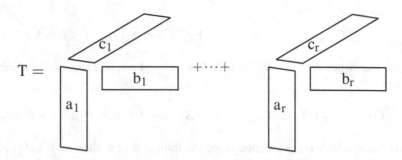

Note that $\mathrm{rank}_{\mathbb{F}}(T)$ is considered an index of complexity of a tensor T similar to matrix rank. In fact, it has been studied in the field of arithmetic complexity (see, for example, Strassen 1983).

By the definition of tensor rank, the following lemma holds immediately.

Lemma 1.1 *For tensors T_1 and T_2,*

$$\mathrm{rank}(T_1 + T_2) \leq \mathrm{rank}(T_1) + \mathrm{rank}(T_2).$$

We also remark that any nonzero 1-tensor, i.e., a nonzero vector, has rank 1.

Next, we describe the properties of rank for a K-mode tensor. Here, we note that

$$V_1 \otimes \cdots \otimes V_K = (V_1 \otimes \cdots \otimes V_L) \otimes (V_{L+1} \otimes \cdots \otimes V_K) \text{ for any } L \text{ with } 1 \leq L \leq K-1.$$

We also note the following lemma.

Lemma 1.2 *Let $v \in V_1 \otimes \cdots \otimes V_L$, $w \in V_{L+1} \otimes \cdots \otimes V_K$. Then, $v \otimes w$ is of rank 1 if and only if v is of rank 1 in the space $V_1 \otimes \cdots \otimes V_L$ and w is of rank 1 in the space $V_{L+1} \otimes \cdots \otimes V_K$.*

Proof This is due to the definition of a rank-1 tensor.

Now, we prove the following.

Lemma 1.3 *Suppose that $v \in V_1 \otimes \cdots \otimes V_L$ and $w \in V_{L+1} \otimes \cdots \otimes V_K$, $\mathrm{rank}(v) = a$ and $\mathrm{rank}(w) = b$, as an element of $V_1 \otimes \cdots \otimes V_L$ and $V_{L+1} \otimes \cdots \otimes V_K$, respectively. Then, $\mathrm{rank}(v \otimes w) \leq ab$.*

Proof Write $v = \sum_{i=1}^{a} v_{i1} \otimes \cdots \otimes v_{iL}$ and $w = \sum_{j=1}^{b} w_{j(L+1)} \otimes \cdots \otimes w_{jK}$. Then,

$$v \otimes w = \sum_{i=1}^{a} \sum_{j=1}^{b} v_{i1} \otimes \cdots \otimes v_{iL} \otimes w_{j(L+1)} \otimes \cdots \otimes w_{jK}.$$

The assertion holds from this equation.

Corollary 1.1 *For $x \in V_1 \otimes \cdots \otimes V_K$, let $x = \sum_{i=1}^{p} y_i \otimes z_i$, where $y_i \in V_1 \otimes \cdots \otimes V_{K-1}$ and $z_i \in V_K$ for $1 \le i \le p$. Suppose that $y_i \in \langle v_1, \ldots, v_r \rangle$ with $\mathrm{rank}(v_i) = 1$ for $1 \le i \le p$, where $\langle v_1, \ldots, v_r \rangle$ is the vector space of $V_1 \otimes \cdots \otimes V_{K-1}$ generated by $v_1, \ldots, v_r$. Then,*

$$\mathrm{rank}(x) \le r. \tag{1.1.1}$$

Proof This is merely a corollary of Lemma 1.3; however, we repeat the proof for the readers' convenience. Set $y_i = \sum_{j=1}^{r} c_{ij} v_j$ for $1 \le i \le p$, where $c_{ij} \in \mathbb{F}$. Then,

$$
\begin{aligned}
x &= \sum_{i=1}^{p} y_i \otimes z_i \\
&= \sum_{i=1}^{p} \left(\sum_{J=1}^{r} c_{ij} v_j \right) \otimes z_i \\
&= \sum_{j=1}^{r} v_j \otimes \left(\sum_{i=1}^{p} c_{ij} z_i \right).
\end{aligned}
$$

Since $\sum_{i=1}^{p} c_{ij} z_i \in V_K$ is of rank 1, from Lemma 1.2, this means that $\mathrm{rank}(x) \le r$.

Corollary 1.2 *Let $x \in V_1 \otimes \cdots \otimes V_K$ and $x = \sum_{j=1}^{N_K} y_j \otimes v_{K,j}$, where $y_j \in V_1 \otimes \cdots \otimes V_{K-1}$ for $1 \le j \le N_K$. Set $\Omega = \{r | v_1, \ldots, v_r \in V_1 \otimes \cdots \otimes V_{K-1} \text{ such that } \mathrm{rank}(v_i) = 1$ for $1 \le i \le r$ and $y_j \in \langle v_1, \ldots, v_r \rangle$ for $1 \le j \le N_K\}$. Then, $\mathrm{rank}(x) = \min \Omega$.*

Proof Since $V_1 \otimes \cdots \otimes V_{K-1}$ is spanned by $\{v_{1i_1} \otimes \cdots \otimes v_{(K-1)i_{(K-1)}} | 1 \le j \le N_j, 1 \le j \le K-1\}$, and $\mathrm{rank}(v_{1i_1} \otimes \cdots \otimes v_{(K-1)i_{(K-1)}}) = 1$ for any $i_1, \ldots, i_{(K-1)}$, we see that the set of Ω is not empty. Let r be the minimum integer of Ω. Then, $\mathrm{rank}(x) \le r$ by definition. Suppose that $\mathrm{rank}(x) \le s < r$. Then, we have

$$x = \sum_{i=1}^{s} y_{1i} \otimes \cdots \otimes y_{(K-1)i} \otimes y_{Ki},$$

where $z_i = y_{1i} \otimes \cdots \otimes y_{(K-1)i}$ is of rank 1 by Lemma 1.2, and therefore, $s \in \Omega$. This is a contradiction.

Further, we have the following.

Proposition 1.1

$$\mathrm{rank}(T) \le \mathrm{max.rank}(N_1, \ldots, N_j) \left(\prod_{j=k+1}^{K} N_u \right).$$

Proof Let us denote $\mathrm{max.rank}(N_1, \ldots, N_j)$ as max.rank here. By assumption, any T has the expression

$$T = \sum_{i_1}^{N_1} \cdots \sum_{i_j=1}^{N_j} \sum_{i_{j+1}=1}^{N_{j+1}} v_{i_1} \otimes \cdots \otimes v_{i_j} \otimes v_{i_{j+1}} \otimes \cdots \otimes v_{i_K}$$

$$= \sum_{i_{j+1}=1}^{N_{j+1}} \cdots \sum_{i_K=1}^{N_K} \sum_{i_1}^{N_1} \cdots \sum_{i_j=1}^{N_j} v_{i_1} \otimes \cdots \otimes v_{i_j} \otimes v_{i_{j+1}} \otimes \cdots \otimes v_{i_K},$$

$$= \sum_{i_{j+1}=1}^{N_{j+1}} \cdots \sum_{i_K=1}^{N_K} \sum_{k=1}^{\text{max.rank}} b_{1k} \otimes \cdots \otimes b_{jk} \otimes v_{i_{j+1}} \otimes \cdots \otimes v_{i_K},$$

$$= \sum_{k=1}^{\text{max.rank}} \sum_{i_{j+1}=1}^{N_{j+1}} \cdots \sum_{i_K=1}^{N_K} b_{1k} \otimes \cdots \otimes b_{jk} \otimes v_{i_{j+1}} \otimes \cdots \otimes v_{i_K}.$$

This proves the assertion.

As a corollary, we have the following.

Proposition 1.2 *For $T \in V_1 \otimes \cdots \otimes V_K$ and $1 \leq j \leq K$,*

$$\text{rank}(T) \leq \left(\prod_{u=1}^{j-1} N_u \right) \left(\prod_{u=j+1}^{K} N_u \right). \tag{1.1.2}$$

Proof It suffices to use Proposition 1.1 inductively.

Since any nonzero vector has rank 1, we also see the following.

Proposition 1.3 *Let T be a 2-mode tensor, i.e., a matrix with column vectors $v_1, \ldots, v_{N_2}$. Then, $\text{rank}(T) = dim\langle v_1, \ldots, v_{N_2} \rangle$, i.e., the tensor rank of T is the same as the one defined in linear algebra.*

1.2   Kronecker Product

Here, we review the Kronecker product between matrices A and B. The Kronecker product might be more familiar than the tensor product to researchers in the field of statistics.

Definition 1.7 For matrices $A = (a_{ij})$ and B, the Kronecker product $\otimes_{kr}$ of A and B is defined by

$$A \otimes_{kr} B = (a_{ij}B). \tag{1.2.1}$$

Note that $A \otimes_{kr} B$ is an $m_1 m_2 \times n_1 n_2$ matrix if A and B are an $m_1 \times n_1$ matrix and an $m_2 \times n_2$ matrix, respectively. The Kronecker product can also be defined between a matrix and a vector or between two vectors.

Remark 1.2 Usually, $\otimes_{kr}$ is simply denoted as $\otimes$. However, we use a new symbol $\otimes_{kr}$ to avoid confusion with the tensor product.

The following holds.

Proposition 1.4 (The fundamental properties of $\otimes_{kr}$)

(1) $(A + B) \otimes_{kr} C = A \otimes_{kr} C + B \otimes_{kr} C$, and $A \otimes_{kr} (B + C) = A \otimes_{kr} B + A \otimes_{kr} C$.
(2) $(A \otimes_{kr} B)^T = A^T \otimes_{kr} B^T$.
(3) $(A \otimes_{kr} B)^{-1} = A^{-1} \otimes_{kr} B^{-1}$ *if A and B are nonsingular.*
(4) $(A \otimes_{kr} B)(C \otimes_{kr} D) = AC \otimes_{kr} BD$ *if AC and BD are definable.*

1.3 *Vec and Tens*

Here, for a tensor T, we define the *vec* (vectorization) operator and its inverse (tensorization) operator *tens* for later use.

Definition 1.8 *vec* is defined as the linear map from $V_1 \otimes \cdots \otimes V_K$ to $\mathbb{F}^{N_1 \cdots N_K}$ such that

$$vec(a_1 \otimes \cdots \otimes a_K) = a_1 \otimes_{kr} \cdots \otimes_{kr} a_K. \tag{1.3.1}$$

This map is an isomorphism, and the inverse map is called a tensorization of a vector v, denoted by $tens(v)$.

Example 1.3 Let $T = (T_{ijk})$ be a $2 \times 2 \times 2$ tensor and $T_{111} = 1$, $T_{121} = 2$, $T_{211} = 3$, $T_{221} = 4$, $T_{112} = 5$, $T_{122} = 6$, $T_{212} = 7$, $T_{222} = 8$. Then, $vec(T) = (1, 2, 3, 4, 5, 6, 7, 8)^T$, and conversely, $tens(vec(T)) = T$. Note that T has a slice representation along the third axis such that

$$T = \left(\begin{pmatrix} 1 & 2 \\ 3 & 4 \end{pmatrix} ; \begin{pmatrix} 5 & 6 \\ 7 & 8 \end{pmatrix} \right).$$

The following proposition holds.

Proposition 1.5 *Let $v_1 \in \mathbb{F}^{N_1}, v_2 \in \mathbb{F}^{N_2}, \ldots, v_K \in \mathbb{F}^{N_K}$ and $A_1, A_2, \ldots, A_K$ be matrices of size $M_1 \times N_1, M_2 \times N_2, \ldots, M_K \times N_K$, respectively. Then,*

$$(A_1 \otimes_{kr} A_2 \otimes_{kr} \cdots \otimes_{kr} A_K)vec(v_1 \otimes v_2 \otimes \cdots \otimes v_K) = vec(A_1 v_1 \otimes A_2 v_2 \otimes \cdots \otimes A_K v_K).$$

Proof From the definition of *vec* and property 4 of Proposition 1.4,

$$
\begin{aligned}
vec(A_1 v_1 \otimes A_2 v_2 \otimes \cdots \otimes A_K v_K) &= A_1 v_1 \otimes_{kr} A_2 v_2 \otimes_{kr} \cdots \otimes_{kr} A_K v_K \\
&= (A_1 \otimes_{kr} A_2 \otimes_{kr} \cdots \otimes_{kr} A_K)(v_1 \otimes_{kr} v_2 \otimes_{kr} \cdots \otimes_{kr} v_K) \\
&= (A_1 \otimes_{kr} A_2 \otimes_{kr} \cdots \otimes_{kr} A_K)vec(v_1 \otimes v_2 \otimes \cdots \otimes v_K).
\end{aligned}
$$

This proves the assertion.

Let A be an $m \times n$ matrix. By abuse of notation, we denote by A the linear map $\mathbb{F}^n \to \mathbb{F}^m; v \to Av$. Let A_i be $M_i \times N_i$ matrices for $1 \le i \le K$. Set $W_i = \mathbb{F}^{M_i}$. Then, $A_1 \otimes_{kr} \cdots \otimes_{kr} A_K$ is a linear map from $V_1 \otimes \cdots \otimes V_K$ to $W_1 \otimes \cdots \otimes W_K$ defined by

$$(A_1 \otimes \cdots \otimes A_K)(v_1 \otimes \cdots \otimes v_K) = A_1(v_1) \otimes \cdots \otimes A_K(v_K).$$

By Proposition 1.5, we see that the following diagram is commutative:

$$
\begin{array}{ccc}
V_1 \otimes \cdots \otimes V_K & \xrightarrow{A_1 \otimes \cdots \otimes A_K} & W_1 \otimes \cdots \otimes W_K \\
{\scriptstyle vec}\downarrow & & \downarrow{\scriptstyle vec} \\
\mathbb{F}^{N_1 \ldots N_K} & \xrightarrow{A_1 \otimes_{kr} \cdots \otimes_{kr} A_K} & \mathbb{F}^{M_1 \ldots M_K}.
\end{array}
$$

We also note the following fact.

Proposition 1.6

$$\operatorname{rank}((A_1 \otimes \cdots \otimes A_K)T) \le \operatorname{rank}(T) \qquad (1.3.2)$$

for any $T \in V_1 \otimes \cdots \otimes V_K$.

Proof Set $r = \operatorname{rank}(T)$ and $T = \sum_{i=1}^r v_{1i} \otimes \cdots \otimes v_{Ki}$. Then,

$$(A_1 \otimes \cdots \otimes A_K)(T) = \sum_{i=1}^r A_1 v_{1i} \otimes \cdots \otimes A_K v_{Ki}.$$

Thus, $\operatorname{rank}((A_1 \otimes \cdots \otimes A_K)(T)) \le r$.

1.4 Mode Products

Let M be an $m \times N_n$ matrix. For $T \in V_1 \otimes \cdots \otimes V_K$, we define the n-mode product $T \times_n A$ between T and A as follows.

Definition 1.9

$$T \times_n M = (E_{N_1} \otimes \cdots E_{N_{n-1}} \otimes M \otimes E_{N_{n+1}} \otimes \cdots \otimes E_{N_K})(T).$$

By the commutativity of the above diagram, we obtain the following proposition.

Proposition 1.7

(1) $T \times_m M_1 \times_n M_2 = T \times_n M_2 \times_m M_1, \quad (m \ne n).$

(2) $T \times_m M_1 \times_n M_2 = T \times_m (M_2 M_1), \quad (m = n).$

Proof First, we prove (1). If $m < n$, then

$$(E_{N_1} \otimes \cdots \otimes E_{N_{n-1}} \otimes M_2 \otimes E_{N_{n+1}} \otimes \cdots \otimes E_{N_K})(E_{N_1} \otimes \cdots \otimes$$
$$E_{N_{m-1}} \otimes M_1 \otimes E_{N_{m+1}} \otimes \cdots \otimes E_{N_K})$$
$$= (E_{N_1} \otimes \cdots \otimes E_{N_{m-1}} \otimes M_1 \otimes E_{N_{m+1}} \otimes \cdots \otimes E_{N_{n-1}} \otimes M_2 \otimes E_{N_{n+1}} \otimes \cdots \otimes E_{N_K}),$$

and

$$(E_{N_1} \otimes \cdots \otimes E_{N_{m-1}} \otimes M_1 \otimes E_{N_{m+1}} \otimes \cdots \otimes E_{N_K})(E_{N_1} \otimes \cdots \otimes$$
$$E_{N_{n-1}} \otimes M_2 \otimes E_{N_{n+1}} \otimes \cdots \otimes E_{N_K})$$
$$= (E_{N_1} \otimes \cdots \otimes E_{N_{m-1}} \otimes M_1 \otimes E_{N_{m+1}} \otimes \cdots \otimes E_{N_{n-1}} \otimes M_2 \otimes E_{N_{n+1}} \otimes \cdots \otimes E_{N_K}).$$

This proves (1). If $m = n$,

$$(E_{N_1} \otimes \cdots \otimes E_{N_{m-1}} \otimes M_2 \otimes E_{N_{m+1}} \otimes \cdots \otimes E_{N_K})(E_{N_1} \otimes \cdots \otimes$$
$$E_{N_{m-1}} \otimes M_1 \otimes E_{N_{m+1}} \otimes \cdots \otimes EN_K)$$
$$= (E_{N_1} \otimes \cdots \otimes E_{N_{m-1}} \otimes M_2 M_1 \otimes E_{N_{m+1}} \otimes \cdots \otimes E_{N_K}).$$

This proves (2).

1.5 Invariance

Let GL(N) denote the set of nonsingular $N \times N$ matrices. Suppose that $M_i \in$ GL(N$_i$) for $1 \le i \le K$. Then, $M_1 \otimes \cdots \otimes M_K$ is an endomorphism of $V_1 \otimes \cdots \otimes V_K$. Since $M_1^{-1} \otimes \cdots \otimes M_K^{-1}$ is the inverse of $M_1 \otimes \cdots \otimes M_K$, we see that $M_1 \otimes \cdots \otimes M_K$ is an automorphism of $V_1 \otimes \cdots \otimes V_K$. The following invariance property of tensor rank holds, which is quite important.

Proposition 1.8 *Set* $g = M_1 \otimes \cdots \otimes M_K$. *Then,*

$$\mathrm{rank}(gT) = \mathrm{rank}(T). \tag{1.5.1}$$

Proof By Proposition 1.6,
$$\mathrm{rank}(gT) \le \mathrm{rank}(T)$$

holds, and for the same reason,

$$\mathrm{rank}(T) = \mathrm{rank}(g^{-1}gT) \le \mathrm{rank}(gT)$$

holds. Thus, $\mathrm{rank}(T) = \mathrm{rank}(gT)$, which proves the assertion.

1.6 Flattening of Tensors

There are several ways to matricize a tensor, referred to as flattening, by which we can determine some properties of the tensor. In this section, we define a type of flattening following Kolda and Bader (2009). In Chap. 2, we will define another type of flattening.

Definition 1.10 Let $f_i : V_1 \otimes \cdots \otimes V_K \to V_i \otimes \mathbb{F}^{N_1 \ldots \hat{N}_i \ldots N_K}$ be a linear map such that

$$f_i(a_1 \otimes \cdots \otimes a_K) = a_i \otimes (a_K \otimes_{kr} \cdots \otimes_{kr} a_{i+1} \otimes_{kr} a_{i-1} \otimes_{kr} \cdots \otimes_{kr} a_1). \quad (1.6.1)$$

Remark 1.3 We identify an element of a tensor product of two vector spaces with a matrix. By this identification,

$$f_i(a_1 \otimes \cdots \otimes a_K) = a_i(a_K \otimes_{kr} \cdots \otimes_{kr} a_{i+1} \otimes_{kr} a_{i-1} \otimes_{kr} \cdots \otimes_{kr} a_1)^T. \quad (1.6.2)$$

Before proceeding, we note that

$$V_1 \otimes \cdots \otimes V_K - (V_1 \otimes \cdots \otimes V_j) \otimes (V_{j+1} \otimes \cdots \otimes V_K).$$

In particular, since $V_1 \otimes \cdots \otimes V_K = V_i \otimes V_1 \otimes \cdots \otimes \hat{V}_i \otimes \cdots \otimes V_K$ for any i, the following hold.

Proposition 1.9
$$\text{rank}(T_{(i)}) \leq \text{rank}(T), \quad (1.6.3)$$

i.e.,
$$\text{max.rank}\{T_{(i)} | 1 \leq i \leq K\} \leq \text{rank}(T). \quad (1.6.4)$$

Proposition 1.10
$$(T \times_i M)_{(i)} = MT_{(i)}. \quad (1.6.5)$$

Proof This follows from the commutativity of the following diagram.

$$
\begin{array}{ccc}
V_1 \otimes \cdots \otimes V_K & \xrightarrow{E_1 \otimes \cdots \otimes E_{N_{i-1}} \otimes M \otimes E_{N_{i+1}} \otimes \cdots \otimes E_K} & V_1 \otimes \cdots \otimes V_K \\
\downarrow & & \downarrow \\
V_i \otimes \mathbb{F}^{N_1 \ldots \hat{N}_i \ldots N_K} & \xrightarrow{M \otimes E} & V_i \otimes \mathbb{F}^{N_1 \ldots \hat{N}_i \ldots N_K}.
\end{array}
$$

Chapter 2
3-Tensors

We summarize several concepts for 3-tensors in this chapter.

2.1 New Flattenings of 3-Tensors

Now, let $T = (T_{ijk})$ be an $N_1 \times N_2 \times N_3$ tensor. We denote $T = (T_1; T_2; \ldots; T_{N_3})$, where T_k are $N_1 \times N_2$ matrices defined by $T_k = (T_{ijk} | 1 \leq i \leq N_1, 1 \leq j \leq N_2)$, $1 \leq k \leq K$. Here, for 3-tensors, we define other flattenings besides the one given in Sect. 1.6.

Definition 2.1 For $T = (T_1; T_2; \ldots; T_{N_3})$ where T_k is an $N_1 \times N_2$ matrix, we set

$$\mathrm{fl}_1(T) = (T_1, T_2, \ldots, T_{N_3}), \qquad (2.1.1)$$

which is an $N_1 \times N_2 N_3$ matrix,
and

$$\mathrm{fl}_2(T) = \begin{pmatrix} T_1 \\ T_2 \\ \vdots \\ T_{N_3} \end{pmatrix}. \qquad (2.1.2)$$

We also provide the following definition.

Definition 2.2 For an $m \times N_1$ matrix P and an $N_2 \times n$ matrix Q, we set

$$PTQ = (PT_1Q, PT_{N_2}Q, PT_{N_3}Q).$$

Remark 2.1

$$PTQ = (P \otimes Q^T \otimes E_{N_3})T.$$

© The Author(s) 2016
T. Sakata et al., *Algebraic and Computational Aspects of Real Tensor Ranks*,
JSS Research Series in Statistics, DOI 10.1007/978-4-431-55459-2_2

By Proposition 1.8, if P and Q are nonsingular matrices, then

$$\text{rank}(PTQ) = \text{rank}(T). \tag{2.1.3}$$

2.2 Characterization of Tensor Rank

Next, we present a characterization of tensor rank through a joint digitalization. Let $T = (A_1; \ldots; A_p)$ be a 3-tensor.

Proposition 2.1 *The rank of T is less than or equal to r if and only if there are an $m \times r$ matrix P, $r \times r$ diagonal matrices D_i, and an $r \times n$ matrix Q such that $A_1 = PD_1Q, \ldots, A_p = PD_pQ$.*

Proof Assume that $\text{rank}(T) \leq r$ and set $T = a_1 \otimes b_1 \otimes c_1 + \cdots + a_r \otimes b_r \otimes c_r$. Further, set $c_i = (c_{i1}, \ldots, c_{ip})^T$ for $1 \leq i \leq r$. Then, $A_k = \sum_{j=1}^{r} c_{ik}B_i$ with $B_i = a_ib_i^T$. Since $T = (A_1; \ldots; A_p)$, if we set $P = (a_1, \ldots, a_r)$, $Q = \begin{bmatrix} b_1^T \\ \vdots \\ b_r^T \end{bmatrix}$, and $D_k = \text{Diag}(c_{1k}, c_{2k}, \ldots, c_{rk})$,

$$PD_kQ = A_k, 1 \leq k \leq p.$$

Conversely, suppose that there are an $m \times r$ matrix P, an $r \times r$ diagonal matrix D_k for $1 \leq k \leq p$, and an $r \times n$ matrix Q such that

$$PD_kQ = A_k, 1 \leq k \leq p.$$

Set $P = (a_1, \ldots, a_r)$, $Q = \begin{bmatrix} b_1^T \\ \vdots \\ b_r^T \end{bmatrix}$, and $D_k = \text{Diag}(c_{1k}, c_{2k}, \ldots, c_{rk})$, $1 \leq k \leq p$. Further, set $c_i = (c_{i1}, \ldots, c_{ip})^T$ for $1 \leq i \leq r$. Since $\sum_{i=1}^{r} c_{ik}a_ib_i^T = A_k$, $1 \leq k \leq p$, $T = \sum_{i=1}^{r} a_i \otimes b_i \otimes c_i$, which completes the proof.

Thus, we have the following.

Proposition 2.2 *The rank of T is the minimum integer r such that there are an $m \times r$ matrix P, $r \times r$ diagonal matrices D_i ($1 \leq i \leq p$), and an $r \times n$ matrix Q such that $A_1 = PD_1Q, \ldots, A_p = PD_pQ$.*

We have the following corollary, and the proof is trivial and omitted.

Corollary 2.1 *Let $T = (A_1; A_2; \ldots; A_p)$ be a $m \times n \times p$ tensor. Let $T_{sub} = (A_1; A_2; \ldots; A_k)$ with $k \leq p$. Then $\text{rank}(T_{sub}) \leq \text{rank}(T)$.*

The next proposition is deduced from Proposition 1.6 in Chap. 1.

Proposition 2.3 *Let $T \in T(m, n, p)$ be expressed as $T = (A_1; A_2; \ldots; A_p)$. Let $r = \max\{\mathrm{rank}(c_1 A_1 + \cdots + c_p A_p) | c_1, \ldots, c_p \in \mathbb{F}\}$. Then, $\mathrm{rank}(T) \geq r$.*

Proof Set $C = \mathrm{Diag}(c_1, \ldots, c_p)$. Then, by Proposition 1.6, it holds that

$$\mathrm{rank}((E_m \otimes E_n \otimes C)T) \leq \mathrm{rank}(T). \tag{2.2.1}$$

Since $(E_m \otimes E_n \otimes C)T = c_1 A_1 + \cdots + c_p A_p$, the result follows.

In general, proving that T has a rank larger than r is not as easy as proving that T has a rank less than or equal to r. The following proposition represents such a case.

Proposition 2.4 *Let $T \in T(n, (m-1)n, m)$ be expressed as m slices of $n \times (m-1)n$ matrices, i.e., $T = (A_1; A_2; \ldots; A_m)$, where $A_1 = (E_n, 0_{n \times (m-1)n})$, $A_2 = (0_{n \times n}, E_n, 0_{n \times n(m-2)}), \ldots, A_{m-1} = (0_{n,(m-2)n}, E_n)$, $A_m = (Y_1, Y_2, \ldots, Y_{m-1})$. Assume that any nontrivial linear combination of $Y_1, \ldots, Y_{m-1}, E_n$ is nonsingular. Then, $\mathrm{rank}(T) > p = (m-1)n$.*

Proof First, note that $\mathrm{fl}_2(T)^{\leq p} = E_p$, where $M^{\leq p}$ denotes the upper p rows of the matrix M. Therefore, we see that $\mathrm{rank}(T) \geq \mathrm{rank}(\mathrm{fl}_2(T)^{\leq p}) = p$. Then, by Proposition 2.1, there are an $m \times p$ matrix P, $p \times p$ diagonal matrices D_k for $1 \leq k \leq m$, and a $p \times p$ matrix Q such that $PD_k Q = A_k$ for $1 \leq k \leq m$. Since

$$E_p = \begin{bmatrix} A_1 \\ A_2 \\ \vdots \\ A_{m-1} \end{bmatrix} = \begin{bmatrix} PD_1 \\ PD_2 \\ \vdots \\ PD_{m-1} \end{bmatrix} Q,$$

Q is nonsingular and

$$Q^{-1} = \begin{bmatrix} PD_1 \\ PD_2 \\ \vdots \\ PD_{m-1} \end{bmatrix}.$$

Set $C = Q^{-1}$. Then, $A_k C = PD_k$ for $1 \leq k \leq m$. In particular, $A_m C = PD_m$. Since $A_m = (Y_1, \ldots, Y_{m-1})$, it holds that

$$Y_1 PD_1 + \cdots + Y_{m-1} PD_{m-1} - PD_m = 0. \tag{2.2.2}$$

Put $D_i = \mathrm{Diag}(d_{i1}, d_{i2}, \ldots, d_{ip})$ for $1 \leq k \leq p$ and $P = (u_1, \ldots, u_p)$. From the Eq. (2.2.2), for the jth column $u_j \neq 0$ of C, it holds that

$$(d_{1,j} Y_1 + d_{2,j} Y_2 + \cdots + d_{m-1,j} Y_{m-1} - d_{m,j} E_n) u_j = 0,$$

which contradicts the assumption that any nontrivial linear combination of $Y_1, \ldots, Y_{m-1}, E_n$ is nonsingular. Thus, $\mathrm{rank}(T) > p = (m-1)n$.

Further, it is quite difficult to show that a tensor has a given rank. However, for some cases, it is possible. Here, we introduce a criterion given by Bi for square-type tensors (Bi 2008).

Proposition 2.5 *Let $T = (A_1, \ldots, A_p)$ be an $n \times n \times p$ tensor, where A_1 is nonsingular. Then, T has rank n if and only if $\{A_j A_1^{-1}, j = 2, \ldots, p\}$ can be diagonalized simultaneously.*

Proof We first prove the "if" part. Since $\{A_j A_1^{-1}, j = 2, \ldots, p\}$ are simultaneously diagonalizable, there is a nonsingular $n \times n$ matrix P such that $P A_j A_1^{-1} P^{-1} = D_j$, where D_j is a diagonal matrix for $2 \leq j \leq p$. $T' = (E_n; A_2 A_1^{-1}; \ldots; A_p A_1^{-1})$ and $T'' = (E_n; D_2; \ldots; D_p)$ is equivalent to T and $\text{rank}(T'') \leq n$. By Proposition 1.9, $\text{rank}(T) \leq n$. By Proposition 1.9, $\text{rank}(T'') \geq n$ and therefore $\text{rank}(T) = \text{rank}(T'') = n$. Next, we prove the "only if" part. Assume that $\text{rank}(T) = n$. By Proposition 2.2, there are $n \times n$ matrices P and Q and $n \times n$ diagonal matrices D_k for $1 \leq k \leq p$. Since A_1 is nonsingular, we see that P, D_1, and Q are nonsingular, and $A_1^{-1} = Q^{-1} D_1^{-1} P^{-1}$. Therefore, $A_j A_1^{-1} = P D_j Q Q^{-1} D_1^{-1} P^{-1} = P D_j D_1^{-1} P^{-1}$ for $2 \leq j \leq p$. Thus, $A_j A_1^{-1}$ are jointly diagonalizable. This proves the assertion.

From this, we have as a special case.

Corollary 2.2 *Let $T = (E_n; A)$ be a tensor of size $(n, n, 2)$. Then, $\text{rank}(T) = n$ if and only if A is diagonalizable.*

Next, we present a condition that a rectangle-type $m \times n \times p$ tensor T has rank p (see Sumi et al. 2015a).

Proposition 2.6 *Let $T \in T(n, (m-1)n, m)$ be expressed as m slices of $n \times (m-1)n$ matrices, i.e., $T = (A_1; A_2; \ldots; A_m)$, where $A_1 = (E_n, 0_{n \times (m-1)n})$, $A_2 = (0_{n \times n}, E_n, 0_{n \times n(m-2)})$, $\ldots$, $A_{m-1} = (0_{n,(m-2)n}, E_n)$, $A_m = (Y_1, Y_2, \ldots, Y_{m-1})$. Then, T has rank p if and only if there are P, $D_i = \text{Diag}(d_{i1}, d_{i2}, \ldots, d_{ip})$, and Q such that $A_i = P D_i Q$, $i = 1, \ldots, m$ and*

$$(d_{1,j} Y_1 + d_{2,j} Y_2 + \cdots + d_{m-1,j} Y_{m-1} - d_{m,j} E_n) u = 0$$

for any column vector u of P.

Proof This is clear from the proof of Proposition 2.4.

2.3 Tensor Rank and Positive Polynomial

For an $n \times n \times p$ tensor T with a slice representation $T = (A_1; A_2; \ldots; A_p)$, we associate T with a polynomial $f_T(x)$, $x \in \mathbb{R}^p$, given by

$$\det{}_T(x) = \det\left(\sum_{i=1}^{p} x_i A_i\right).$$

If this polynomial is positive for all $x = (x_1, \ldots, x_p) \neq 0$, the tensor is called an absolutely nonsingular tensor. Absolutely nonsingular tensors are closely related to the rank determination problem (see Sakata et al. 2011 and Sumi et al. 2010, 2014, 2013). Therefore, we treat it in Chap. 5 in greater detail. Here, we only note that this type of linear combination of matrices has been considered in the system stability problem in engineering fields, and the positivity of multivariate polynomials has been a central topic in the field of algebra with regard to Hilbert's 17th problem.

If this polynomial is positive for all $x = (x_1, \ldots, x_p) \neq 0$, the tensor is called an absolutely nonsingular tensor. Absolutely nonsingular tensors are closely related to the rank determination problem (see Sakata et al. 2011 and Sumi et al. 2010, 2014, 2013). Therefore, we treat it in Chap. 5 in greater detail. Here, we only note that this type of linear combination of matrices has been considered in the system stability problem in engineering fields, and the positivity of multivariate polynomials has been a central topic in the field of algebra with regard to Hilbert's 17th problem. For these topics, for example, see Artin (1927), de Loera and Santos (1996), Lasserre (2010), Powers (2011), Powers and Reznick (2001, 2005), Prestel and Delzell (2001), Rangel (2009), Schmüdgen (1991), Schweighofer (2002), and Stengle (1974). These relations between several fields are quite interesting.

Chapter 3
Simple Evaluation Methods of Tensor Rank

In this chapter, we illustrate simple evaluations of the rank of 3-tensors, which might facilitate readers' understanding of tensor rank. Throughout this section, we consider ranks only over the real filed $\mathbb{R}$, and we abbreviate the symbol $\mathbb{R}$ from all notations. Here we consider the maximal rank of $2 \times 2 \times 2, 2 \times 2 \times 3, 2 \times 3 \times 3$, and $3 \times 3 \times 3$ tensors. Note that "maximal rank" is simply the maximum of the rank in a set of a type of tensors (see Chap. 5 for further details).

3.1 Key Lemma

Throughout this chapter, we use the following key lemma for evaluation of tensor rank.

Lemma 3.1 *Let T and T_1 be $n \times n \times 3$ tensors and $\mathrm{rank}(T_1) = 1$. Let $T_2 = T - T_1 = (A_1; A_2; A_3)$. Then if (1) $A_1^{-1}A_2$ or (2) $A_2^{-1}A_1$ is diagonalizable, $\mathrm{rank}(T) \leq n + 1 + \mathrm{rank}(A_3)$.*

Proof For case (1), since $A_1^{-1}T = A_1^{-1}T_1 + T_2$ and $T_2 = T_{21} + T_{22}$, where $T_{21} = (E_n; A_1^{-1}A_2; 0)$ and $T_{22} = (0; 0; A_1^{-1}A_3)$, $\mathrm{rank}(T) = \mathrm{rank}(A_1^{-1}T) \leq \mathrm{rank}(A_1^{-1}T_1) + \mathrm{rank}(T_{21}) + \mathrm{rank}(T_{22}) \leq 1 + n + \mathrm{rank}(A_1^{-1}A_3) = 1 + n + \mathrm{rank}(A_3)$. Case (2) is proved similarly. This completes the proof.

Remark 3.1 This lemma is applicable to $n \times n \times 2$ tensors by considering $A_3 = 0$.

Remark 3.2 For example,

$$T_1 = \left(\begin{pmatrix} \alpha & 0 \\ 0 & 0 \end{pmatrix} ; \begin{pmatrix} \beta & 0 \\ 0 & 0 \end{pmatrix} \right)$$

and

$$T_1 = \left(\begin{pmatrix} 0 & 0 \\ 0 & \alpha \end{pmatrix} ; \begin{pmatrix} 0 & 0 \\ 0 & \beta \end{pmatrix} \right)$$

T. Sakata et al., *Algebraic and Computational Aspects of Real Tensor Ranks*, JSS Research Series in Statistics, DOI 10.1007/978-4-431-55459-2_3

are $2 \times 2 \times 2$ rank-1 tensors and

$$T_1 = \left(\begin{pmatrix} 0\,0\,0 \\ \alpha\,0\,0 \\ 0\,0\,0 \end{pmatrix} ; \begin{pmatrix} 0\,0\,0 \\ \beta\,0\,0 \\ 0\,0\,0 \end{pmatrix} ; \begin{pmatrix} 0\,0\,0 \\ \gamma\,0\,0 \\ 0\,0\,0 \end{pmatrix} \right)$$

is a rank-1 $3 \times 3 \times 3$ tensor, which are used in the following sections.

3.2 Maximal Rank of $2 \times 2 \times 2$ Tensors

A $2 \times 2 \times 2$ tensor is the smallest tensor. We will present a short proof for the following well-known fact.

Proposition 3.1 *It holds that* max.rank$(2, 2, 2) = 3$.

Lemma 3.2 *(1) Let $A = \begin{pmatrix} \alpha & 0 \\ 0 & 1 \end{pmatrix}$ and $B = (b_1, kb_1)$, where b_1 is a two-dimensional vector, both of whose elements are not zeros. Then, for appropriate α, A is nonsingular and $A^{-1}B$ is diagonalizable, and setting $T = (A; B)$, rank$(T) \le 2$ by Lemma 3.2.*
(2) Let $A = \begin{pmatrix} 1 & 0 \\ 0 & \alpha \end{pmatrix}$ and $B = (kb_2, b_2)$, where b_2 is a two-dimensional vector, both of whose elements are not zeros. Then, for appropriate α, A is nonsingular and $A^{-1}B$ is diagonalizable, and setting $T = (A; B)$, rank$(T) \le 2$ by Lemma 3.2.

Proof Proof of (1). Let $A = \begin{pmatrix} \alpha & 0 \\ 0 & 1 \end{pmatrix}$ and $B = \begin{pmatrix} a & ka \\ c & kc \end{pmatrix}$. Then,

$$A^{-1}B = \frac{1}{\alpha}\begin{pmatrix} 1 & 0 \\ 0 & \alpha \end{pmatrix}\begin{pmatrix} a & ka \\ c & kc \end{pmatrix} = \frac{1}{\alpha}\begin{pmatrix} a & ka \\ \alpha c & \alpha kc \end{pmatrix}$$

and the characteristic polynomial of $A^{-1}B$ is

$$f(x) = \frac{1}{\alpha}x(x - a - \alpha kc).$$

Therefore, for $k \ne 0$, taking $\alpha = -\dfrac{a}{kc}$, $f(x)$ has two different roots and $A^{-1}B$ is diagonalizable. For $k = 0$, $f(x)$ has two roots, 0 and $a \ne 0$, and $A^{-1}B$ is diagonalizable, which implies that rank$(T) \le 2$.

Proof of (2). Let $A = \begin{pmatrix} 1 & 0 \\ 0 & \alpha \end{pmatrix}$ and $B = \begin{pmatrix} ck & c \\ kd & d \end{pmatrix}$. Then

$$A^{-1}B = \frac{1}{\alpha}\begin{pmatrix} \alpha & 0 \\ 0 & 1 \end{pmatrix}\begin{pmatrix} kc & c \\ kd & d \end{pmatrix} = \frac{1}{\alpha}\begin{pmatrix} \alpha kc & \alpha c \\ kd & d \end{pmatrix}$$

and the characteristic polynomial of $A^{-1}B$ is

$$f(x) = \frac{1}{\alpha}x(x - d - \alpha kc).$$

Therefore, for $k \neq 0$, taking $\alpha = -\dfrac{d}{kc}$, $f(x)$ has two different roots and $A^{-1}B$ is diagonalizable. For $k = 0$, f(x) has two roots, 0 and $d \neq 0$, and $A^{-1}B$ is diagonalizable, which implies that rank$(T) \leq 2$.

First, we prove the following:

Proposition 3.2
$$\text{max.rank}(T(2, 2, 2)) \leq 3.$$

Before proceeding to the proof, we note the following fact.

Fact 3.1 *Let $A = \begin{pmatrix} s & t \\ u & w \end{pmatrix}$ be a 2×2 nonsingular matrix. It has seven patterns, as shown below, where $stuw \neq 0$.*

(1) $\begin{array}{|c|c|} \hline s & t \\ \hline u & w \\ \hline \end{array}$, (2) $\begin{array}{|c|c|} \hline 0 & t \\ \hline u & w \\ \hline \end{array}$, (3) $\begin{array}{|c|c|} \hline s & t \\ \hline 0 & w \\ \hline \end{array}$, (4) $\begin{array}{|c|c|} \hline s & t \\ \hline u & 0 \\ \hline \end{array}$, (5) $\begin{array}{|c|c|} \hline s & 0 \\ \hline u & w \\ \hline \end{array}$, (6) $\begin{array}{|c|c|} \hline s & 0 \\ \hline 0 & w \\ \hline \end{array}$, (7) $\begin{array}{|c|c|} \hline 0 & t \\ \hline u & 0 \\ \hline \end{array}$.

Now, we start with the proof. Our proof is given for each of the above seven patterns.

Proof Let $T = (A; B)$, where A and B are 2×2 matrices.
(1) If rank$(A) \leq 1$ and rank$(B) \leq 1$, then rank$(T) \leq 2$.
(2) Without loss of generality, we assume that rank$(A) = 2$. Set $T' = (A^{-1}A; A^{-1}B)$ $= (E_2; B')$. If B' is singular, it is obvious that rank$(T) \leq 2 + 1 = 3$. Therefore, we assume that rank$(B') = 2$. Put $B' = \begin{pmatrix} a & b \\ c & d \end{pmatrix}$. We consider seven cases.

(2-1) $B' = \begin{pmatrix} a & b \\ c & d \end{pmatrix}$, $abcd \neq 0$.

Let $T_1 = \left(\begin{pmatrix} -\alpha + a & 0 \\ 0 & 0 \end{pmatrix}; \begin{pmatrix} \beta & 0 \\ 0 & 0 \end{pmatrix} \right)$. Then rank$(T_1) = 1$.

$$T_2 = T - T_1 = (A_2; B_2) = \left(\begin{pmatrix} \alpha & 0 \\ 0 & 1 \end{pmatrix}; \begin{pmatrix} \beta & b \\ c & d \end{pmatrix} \right). \qquad (3.2.1)$$

Now, we choose β such that B_2 is of rank 1, and therefore, we have

$$T_2 = \left(\begin{pmatrix} \alpha & 0 \\ 0 & 1 \end{pmatrix}; \begin{pmatrix} a & ka \\ c & kc \end{pmatrix} \right).$$

Then, by (1) of Lemma 3.2, rank$(T_2) \leq 2$ and rank$(T) = $ rank$(T') \leq$ rank$(T_1) +$ rank$(T_2) = 1 + 2 = 3$.

(2-2) $B' = \begin{pmatrix} 0 & b \\ c & d \end{pmatrix}$, $bcd \neq 0$.

Let

$$T_1 = \left(\begin{pmatrix} -\alpha + a & 0 \\ 0 & 0 \end{pmatrix}; \begin{pmatrix} \beta & 0 \\ 0 & 0 \end{pmatrix} \right).$$

Then, $\text{rank}(T_1) = 1$.

$$T_2 = T - T_1 = (A_2; B_2) = \left(\begin{pmatrix} \alpha & 0 \\ 0 & 1 \end{pmatrix}; \begin{pmatrix} \beta & b \\ c & d \end{pmatrix} \right).$$

Now, we choose β such that B_2 is of rank 1, and therefore, we have

$$T_2 = \left(\begin{pmatrix} \alpha & 0 \\ 0 & 1 \end{pmatrix}; \begin{pmatrix} a & ka \\ c & kc \end{pmatrix} \right).$$

This reduces to Eq. (3.2.1), and $\text{rank}(T) \leq 3$.

(2-3) $B' = \begin{pmatrix} a & b \\ 0 & d \end{pmatrix}$, $abd \neq 0$.

Let

$$T_1 = (A_1; B_1) = \left(\begin{pmatrix} -\alpha & 0 \\ 0 & 0 \end{pmatrix}; \begin{pmatrix} a & 0 \\ 0 & 0 \end{pmatrix} \right).$$

Then, $\text{rank}(T_1) = 1$. We have

$$T_2 = T - T_1 = (A_2; B_2) = \left(\begin{pmatrix} \alpha & 0 \\ 0 & 1 \end{pmatrix}; \begin{pmatrix} 0 & b \\ 0 & d \end{pmatrix} \right).$$

This is the case of (2) with $k = 0$ in Lemma 3.2, and $\text{rank}(T) \leq 3$.

(2-4) $B' = \begin{pmatrix} a & b \\ c & 0 \end{pmatrix}$, $abc \neq 0$.

Let

$$T_1 = (A_1; B_1) = \left(\begin{pmatrix} 0 & 0 \\ 0 & -\alpha \end{pmatrix}; \begin{pmatrix} 0 & 0 \\ 0 & -\beta \end{pmatrix} \right).$$

Then $\text{rank}(T_1) = 1$. Now, we choose β such that B_2 is of rank 1, and therefore, we have

$$T_2 = T - T_1 = (A_2; B_2) = \left(\begin{pmatrix} 1 & 0 \\ 0 & \alpha \end{pmatrix}; \begin{pmatrix} kb & b \\ kd & d \end{pmatrix} \right).$$

This is the case of (2) in Lemma 3.2, and $\text{rank}(T) \leq 3$.

(2-5) $B' = \begin{pmatrix} a & 0 \\ c & d \end{pmatrix}$, $acd \neq 0$.

Let

$$T_1 = (A_1; B_1) = \left(\begin{pmatrix} 0 & 0 \\ 0 & -\alpha \end{pmatrix}; \begin{pmatrix} 0 & 0 \\ 0 & -\beta \end{pmatrix} \right).$$

Then, $\text{rank}(T_1) = 1$. Now, we choose β such that B_2 is of rank 1, and therefore, we have

$$T_2 = T - T_1 = (A_2; B_2) = \left(\begin{pmatrix} 1 & 0 \\ 0 & \alpha \end{pmatrix}; \begin{pmatrix} a & 0 \\ c & 0 \end{pmatrix} \right).$$

This is the case of (1) with $k = 0$ in Lemma 3.2, and $\text{rank}(T) \leq 3$.

(2-6) $B' = \begin{pmatrix} a & 0 \\ 0 & d \end{pmatrix}$, $ad \neq 0$.

For this case, clearly, we have $\text{rank}(T) \leq 2$.

(2-7) $B' = \begin{pmatrix} 0 & b \\ c & 0 \end{pmatrix}$, $bc \neq 0$.

Let

$$T_1 = (A_1; B_1) = \left(\begin{pmatrix} 0 & -1 \\ -1 & 0 \end{pmatrix}; \begin{pmatrix} 0 & b \\ c & 0 \end{pmatrix} \right).$$

Then, $\text{rank}(T_1) = 1$. Note that B_2 is a null matrix and we have

$$T_2 = T - T_1 = (A_2; B_2) = \left(\begin{pmatrix} 1 & 1 \\ 1 & 1 \end{pmatrix}; \begin{pmatrix} 0 & 0 \\ 0 & 0 \end{pmatrix} \right).$$

Thus, $\text{rank}(T) \leq 1 + 1 + 1 = 3$. This completes the proof.

Next, we prove the following proposition.

Proposition 3.3 *There is a $2 \times 2 \times 2$ tensor T with* $\text{rank}(T) = 3$.

Proof Set $A = \begin{pmatrix} 1 & 0 \\ 0 & 1 \end{pmatrix}$ and $B = \begin{pmatrix} 0 & 0 \\ 1 & 0 \end{pmatrix}$. Since B is not diagonalizable, by Proposition 2.5, $\text{rank}(T) \neq 2$, it is clear that $\text{rank}(T) = 3$. This proves Proposition 3.3. $\qquad \blacksquare$

From Propositions 3.2 and 3.3, Proposition 3.1 holds.

3.3   Maximal Rank of $2 \times 2 \times 3$ and $2 \times 3 \times 3$ Tensors

We prove the following proposition.

Proposition 3.4
$$\text{max.rank}(2, 2, 3) = 3. \tag{3.3.1}$$

Proof Let $T = (A; B)$, where A and B are 2×3 matrices. We denote $A = (a, b, c)$ and $B = (d, e, f)$, where a, b, c, d, e, and f are two-dimensional column vectors. If one of A and B is of rank 1, there is nothing to prove since $\text{rank}(T) \leq \text{rank}((A; 0)) + \text{rank}((0; B)) \leq 1 + 2 \ (2 + 1) \ = 3$. Therefore, we assume that both of A and B are of full rank. Without loss of generality, we assume that $a \perp b$ (linearly independent), and by column operations, we can transform $c = 0$. Then, if f is the zero vector, it is clear that $\mathrm{rank}(T) \leq \max.\mathrm{rank}(2, 2, 2) = 3$. Therefore, we assume that $f \neq (0, 0)$. If both d and e are constant multiples of f, then $\mathrm{rank}(T) \leq 2 + 1$. Therefore, without loss of generality, we assume that $e \perp f$. By a column operation we can assume that $d = 0$. Since $a \perp b$, we can write $e = \alpha a + \beta b$ and $f = \gamma a + \delta b$. If $\gamma \neq 0$, by a column operation, we can make a change $e = \beta b \leftarrow e = \alpha a + \beta b$, i.e., $T = ((a, b, 0); (0, \beta b, \gamma a + \delta b))$, which means that $\mathrm{rank}(T) \leq 3$. Therefore, we assume that $\gamma = 0$ and $T = ((a, b, 0); (0, \alpha a + \beta b, \delta b))$. If $\delta = 0$, $T = ((a, b, 0); (0, \alpha a + \beta b, 0))$, which means that $\mathrm{rank}(T) \leq 1 + 1 + 1 = 3$. If $\delta \neq 0$, by column operations, we have $T = ((a, \alpha a + b, 0); (0; \alpha a + b, \delta b))$, which means that $\mathrm{rank}(T) \leq 3$. Thus, we have proved that $\max.\mathrm{rank}(2, 2, 3) \leq 3$.

Now, Let T be $\left(\left(\begin{smallmatrix} 1 & 0 & 0 \\ 0 & 1 & 0 \end{smallmatrix} \right); \left(\begin{smallmatrix} 0 & 0 & 0 \\ 1 & 0 & 0 \end{smallmatrix} \right) \right)$. Then we can view T as $\left(\left(\begin{smallmatrix} 1 & 0 \\ 0 & 1 \end{smallmatrix} \right); \left(\begin{smallmatrix} 0 & 0 \\ 1 & 0 \end{smallmatrix} \right); \left(\begin{smallmatrix} 0 & 0 \\ 0 & 0 \end{smallmatrix} \right) \right)$. By Corollary 2.1, $\mathrm{rank}(T) \geq \mathrm{rank}(T_{sub}) = 3$, where $T_{sub} = \left(\left(\begin{smallmatrix} 1 & 0 \\ 0 & 1 \end{smallmatrix} \right); \left(\begin{smallmatrix} 0 & 0 \\ 1 & 0 \end{smallmatrix} \right) \right)$. This completes the proof.

Next, we prove the following proposition.

Proposition 3.5

$$\max.\mathrm{rank}(2, 3, 3) = 4 \qquad\qquad (3.3.2)$$

Proof Let $T = (A_1; A_2)$, where A_1 and A_2 are 3×3 matrices. If both A_1 and A_2 are of rank 2, then we have nothing to prove. Without loss of generality, assume that A_1 is nonsingular. Considering an exchange $A_2 \leftarrow x_0 A_1 + A_2$ with some appropriate x_0, we can assume that A_2 is of rank 2. Then by column and row operations we have

$$T = \left(\begin{pmatrix} a_{11} & a_{12} & a_{13} \\ a_{21} & a_{22} & a_{23} \\ a_{31} & a_{32} & a_{33} \end{pmatrix}; \begin{pmatrix} b_{11} & b_{12} & 0 \\ a_{21} & a_{22} & 0 \\ 0 & 0 & 0 \end{pmatrix} \right).$$

From this expression, it is easy to see that $\mathrm{rank}(T) \leq \max.\mathrm{rank}(2, 2, 3) + 1 = 3 + 1 = 4$.

Next, we set

$$T = (A_1; A_2) = \left(\begin{pmatrix} 1 & 0 & 0 \\ 0 & 1 & 0 \\ 0 & 0 & 1 \end{pmatrix}; \begin{pmatrix} 0 & 0 & 0 \\ 0 & 1 & 0 \\ 1 & 0 & 0 \end{pmatrix} \right).$$

From Lemma 2.5, T is of rank 3 if and only if A_2 is diagonalizable. Since A_2 is not diagonalizable, $\mathrm{rank}(T) \neq 3$, which implies that $\mathrm{rank}(T) = 4$. This completes the proof.

3.4 Maximal Rank of $3 \times 3 \times 3$ Tensors

In this section, we use the notation $RC(i \leftrightarrow j)$ for the exchanges between the ith row and the jth row and the ith column and the jth column. Further, by $A \leftarrow B$, we denote the exchange of a matrix A by a matrix A', and the changed $A(A')$ is sometimes denoted by the same symbol A for notational simplicity.

The following is known.

Proposition 3.6 *The real maximal rank of* $3 \times 3 \times 3$ *tensors is equal to 5.*

Although this is a well-known fact, it requires a quite delicate argument (see, for example, Sumi et al. 2010). Here, we give a detailed proof by simple linear algebraic methods with a complete detailed decomposition into patterns of a $3 \times 3 \times 3$ tensor. First, we prove the following.

Proposition 3.7

$$\text{max.rank}(T) = 5 \text{ for } T \in T(3, 3, 3). \tag{3.4.1}$$

Proof Let $T = (A_1; A_2; A_3)$, where A_1, A_2, A_3 are 3×3 matrices. If some of A_1, A_2, and A_3 are singular, we exchange among them and we can assume that A_3 is singular. If all of A_1, A_2, and A_3 are nonsingular, we consider a polynomial $f(x) = |xA_1 + A_3| = |A_1||xE + A_1^{-1}A_3|$, which is a real polynomial with degree 3. Therefore, $f(x)$ vanishes at some x_0. Thus, $x_0A_1 + A_3$ is singular and we exchange $A_3 \leftarrow A_3 + x_0A_1$. Thus, A_3 can be assumed to be singular. Then, we transform T into

$$T = \left(\begin{pmatrix} a_{11} & a_{12} & a_{13} \\ a_{21} & a_{22} & a_{23} \\ a_{31} & a_{32} & a_{33} \end{pmatrix} ; \begin{pmatrix} b_{11} & b_{12} & b_{13} \\ b_{21} & b_{22} & b_{23} \\ b_{31} & b_{32} & b_{33} \end{pmatrix} ; \begin{pmatrix} * & * & 0 \\ * & * & 0 \\ 0 & 0 & 0 \end{pmatrix} \right).$$

If $a_{33} \neq 0$, by column and row operations, T becomes

$$T = \left(\begin{pmatrix} a_{11} & a_{12} & 0 \\ a_{21} & a_{22} & 0 \\ 0 & 0 & a_{33} \end{pmatrix} ; \begin{pmatrix} b_{11} & b_{12} & b_{13} \\ b_{21} & b_{22} & b_{23} \\ b_{31} & b_{32} & b_{33} \end{pmatrix} ; \begin{pmatrix} * & * & 0 \\ * & * & 0 \\ 0 & 0 & 0 \end{pmatrix} \right).$$

Then, if necessary, exchange $A_2 \leftarrow kA_1 + A_2$ with an appropriate k, and we assume that $b_{33} \neq 0$. If we let

$$T_1 = \left(\begin{pmatrix} 0 & 0 & 0 \\ 0 & 0 & 0 \\ 0 & 0 & a_{33} \end{pmatrix} ; \begin{pmatrix} 0 & 0 & 0 \\ 0 & 0 & 0 \\ 0 & 0 & 0 \end{pmatrix} ; \begin{pmatrix} 0 & 0 & 0 \\ 0 & 0 & 0 \\ 0 & 0 & 0 \end{pmatrix} \right),$$

$T_2 = T - T_1$ is given by

$$T_2 = \left(\begin{pmatrix} a_{11} & a_{12} & 0 \\ a_{21} & a_{22} & 0 \\ 0 & 0 & 0 \end{pmatrix} ; \begin{pmatrix} b_{11} & b_{12} & b_{13} \\ b_{21} & b_{22} & b_{23} \\ b_{31} & b_{32} & b_{33} \end{pmatrix} ; \begin{pmatrix} * & * & 0 \\ * & * & 0 \\ 0 & 0 & 0 \end{pmatrix} \right).$$

Then by column and row operations, T_2 becomes

$$T_2 = \left(\begin{pmatrix} a_{11} & a_{12} & 0 \\ a_{21} & a_{22} & 0 \\ 0 & 0 & 0 \end{pmatrix} ; \begin{pmatrix} b_{11} & b_{12} & 0 \\ b_{21} & b_{22} & 0 \\ 0 & 0 & b_{33} \end{pmatrix} ; \begin{pmatrix} * & * & 0 \\ * & * & 0 \\ 0 & 0 & 0 \end{pmatrix} \right).$$

Further, if we let

$$T_3 = \left(\begin{pmatrix} 0 & 0 & 0 \\ 0 & 0 & 0 \\ 0 & 0 & 0 \end{pmatrix} ; \begin{pmatrix} 0 & 0 & 0 \\ 0 & 0 & 0 \\ 0 & 0 & b33 \end{pmatrix} ; \begin{pmatrix} 0 & 0 & 0 \\ 0 & 0 & 0 \\ 0 & 0 & 0 \end{pmatrix} \right),$$

$T_2 - T_3$ becomes

$$T_3 = \left(\begin{pmatrix} a_{11} & a_{22} & 0 \\ a_{21} & a_{22} & 0 \\ 0 & 0 & 0 \end{pmatrix} ; \begin{pmatrix} b_1 & b_{12} & 0 \\ b_{21} & b_{22} & 0 \\ 0 & 0 & 0 \end{pmatrix} ; \begin{pmatrix} * & * & 0 \\ * & * & 0 \\ 0 & 0 & 0 \end{pmatrix} \right).$$

From these, we can evaluate that

$$\text{rank}(T) \le \text{rank}(T_1) + \text{rank}(T_2) + \text{max.rank}(2, 2, 2) = 1 + 1 + 3 = 5.$$

Thus, we assume that $a_{33} = 0$. Similarly, we assume that $b_{33} = 0$. Thus, we assume that

$$T = \left(\begin{pmatrix} a_{11} & a_{12} & a_{13} \\ a_{21} & a_{22} & a_{23} \\ a_{31} & a_{32} & 0 \end{pmatrix} ; \begin{pmatrix} b_{11} & b_{12} & b_{13} \\ b_{21} & b_{22} & b_{23} \\ b_{31} & b_{32} & 0 \end{pmatrix} ; \begin{pmatrix} * & * & 0 \\ * & * & 0 \\ 0 & 0 & 0 \end{pmatrix} \right).$$

Next, for the case $a_{31} = a_{32} = 0$, we can evaluate that

$$\text{rank}(T) \le 1 + \text{max.rank}(2, 3, 3) = 1 + 4 = 5.$$

Therefore, we assume that $(a_{31}, a_{32}) \neq (0, 0)$. Similarly, we can assume that $(a_{13}, a_{23}) \neq (0, 0)$. By row and column operations and a constant multiplication, T becomes

$$T = \left(\begin{pmatrix} a_{11} & a_{12} & 1 \\ a_{21} & a_{22} & 0 \\ 1 & 0 & 0 \end{pmatrix} ; \begin{pmatrix} b_{11} & b_{12} & b_{13} \\ b_{21} & b_{22} & b_{23} \\ b_{31} & b_{32} & 0 \end{pmatrix} ; \begin{pmatrix} * & * & 0 \\ * & * & 0 \\ 0 & 0 & 0 \end{pmatrix} \right).$$

Similarly, we can assume that $(b_{31}, b_{32}) \neq (0, 0)$ and $(b_{13}, b_{23}) \neq (0, 0)$. If $(b_{31}, b_{32}) = (b_{31}, 0)$ or $(b_{13}, b_{23}) = (b_{13}, 0)$, we can evaluate that

$$\text{rank}(T) \le 1 + \text{max.rank}(2, 3, 3) = 1 + 4 = 5.$$

Therefore, we can assume that $b_{23} \neq 0$ and $b_{32} \neq 0$. Then by column and row operations and constant multiplications, T becomes

$$T = \left(\begin{pmatrix} a_{11} & a_{12} & 1 \\ a_{21} & a_{22} & 0 \\ 1 & 0 & 0 \end{pmatrix} ; \begin{pmatrix} b_{11} & b_{12} & 0 \\ b_{21} & b_{22} & 1 \\ 0 & 1 & 0 \end{pmatrix} ; \begin{pmatrix} * & * & 0 \\ * & * & 0 \\ 0 & 0 & 0 \end{pmatrix} \right).$$

Furthermore, let $P_1 = \begin{pmatrix} 1 & 0 & 0 \\ 0 & 1 & -a_{21} \\ 0 & 0 & 1 \end{pmatrix}$, $Q_1 = \begin{pmatrix} 1 & 0 & 0 \\ 0 & 1 & 0 \\ 0 & -a_{21} & 1 \end{pmatrix}$, $P_2 = \begin{pmatrix} 1 & 0 & -b_{12} \\ 0 & 1 & 0 \\ 0 & 0 & 1 \end{pmatrix}$,

$Q_2 = \begin{pmatrix} 1 & 0 & 0 \\ 0 & 1 & 0 \\ -b_{21} & 0 & 1 \end{pmatrix}$ and let $P = P_2 P_1$ and $Q = Q_1 Q_2$. Then, PTQ has the form

$$PTQ = \left(\begin{pmatrix} a_{11} & 0 & 1 \\ 0 & a_{22} & 0 \\ 1 & 0 & 0 \end{pmatrix} ; \begin{pmatrix} b_{11} & 0 & 0 \\ 0 & b_{22} & 1 \\ 0 & 1 & 0 \end{pmatrix} ; \begin{pmatrix} * & * & 0 \\ * & * & 0 \\ 0 & 0 & 0 \end{pmatrix} \right). \qquad (3.4.2)$$

Now, we show the inequality (3.4.1) for the canonical form (3.4.2).

We use the seven patterns given in Fact 3.1. Here these patterns are grouped into three groups, $G_i, i = 1, 2, 3$, such that $G_1 = \{(1), (2), (3), (4), (7)\}, G_2 = \{(5)\}$, and $G_3 = \{(6)\}$. The feature of the members of G_1 is that subtraction of an appropriate number from the $(2, 1)$ element makes the matrix a rank-1 matrix. On the other hand, the feature of the elements of G_2 is that they become members of G_1 by $RC(1 \leftrightarrow 2)$. G_3 consists of diagonal matrices.

We begin with the following elementary lemmas.

Lemma 3.3 *Let $f(x)$ be a monic polynomial with degree 3. The following holds.*

(1) If $f(0) > 0$ and $f(x_0) < 0$ at some $x_0 > 0$, then $f(x)$ has three real roots.
(2) If $f(0) < 0$ and $f(x_0) > 0$ at some $x_0 < 0$, then $f(x)$ has three real roots.

Proof This is a straightforward fact; hence, the proof is omitted.

Lemma 3.4 *Let*

$$f(x) = x^3 + \alpha x^2 + \beta x + c.$$

Then, $f(x) = 0$ has three real roots for appropriate α and β.

Proof By assumption, if $f(0) = c > 0$, $f(1) = 1 + \alpha + \beta + c(\alpha, \beta)$ becomes a negative value for appropriate α and β. If $f(0) = c(\alpha, \beta) < 0$, $f(-1) = -1 + \alpha - \beta + c(\alpha, \beta)$ and this becomes a positive value for appropriate α and β. From Lemma 3.3, the assertion holds.

Remark 3.3 In order to prove that a monic polynomial with degree 3 has real three zeros, we can use the discriminant of the polynomial. However, it usually becomes a complicated function of the coefficients, and we avoid using it here.

Now, we start over with the proof of Proposition 3.4.1. Let

$$T = \left(\begin{pmatrix} a_{11} & 0 & 1 \\ 0 & a_{22} & 0 \\ 1 & 0 & 0 \end{pmatrix} ; \begin{pmatrix} b_{11} & 0 & 0 \\ 0 & b_{22} & 1 \\ 0 & 1 & 0 \end{pmatrix} ; \begin{pmatrix} c_{11} & c_{12} & 0 \\ c_{21} & c_{22} & 0 \\ 0 & 0 & 0 \end{pmatrix} \right).$$

Let $M = \begin{pmatrix} c_{11} & c_{12} \\ c_{21} & c_{22} \end{pmatrix}$.

We divide the proofs according to the three patterns of M

(1) The case of $M \in G_1$

(1-1) $a_{22} \neq 0$.

Then A_1 is nonsingular and let $A_1 \leftarrow A_1 + \alpha E_{21}$, $A_2 \leftarrow A_2 + \beta E_{21}$, and $A_3 \leftarrow A_3 + \gamma E_{21}$ with appropriate (α, β, γ). In particular, γ is chosen such that A_3 is of rank 1. Then, let $f(x)$ be the eigen polynomial or characteristic polynomial of $A_1^{-1} A_2$ and $y = -a_{22} x$. Then

$$f(-\frac{y}{a_{22}}) = g(y) = -\frac{1}{a_{22}^3}(y^3 + (b_{22} - \alpha)y^2 + (-\beta a_{22} + a_{22}a_{11})y + a_{22}^2 b_{11}).$$

Since $a_{22} \neq 0$, the equation $g(y) = 0$ and the equation $f(x) = 0$ have three real roots for appropriate α and β from Lemma 3.4, and thus $A_1^{-1} A_2$ is diagonalizable. From Lemma 3.2, this proves that $\text{rank}(T) \leq 5$.

(1-2) $a_{22} = 0$ and $b_{11} \neq 0$.

Let $A_1 \leftarrow A_1 + \alpha E_{21}$, $A_2 \leftarrow A_2 + \beta E_{21}$, and $A_3 \leftarrow A_3 + \gamma E_{21}$ with appropriate (α, β, γ). In particular, γ is chosen such that A_3 is of rank 1. Then, A_2 is nonsingular for any β. Then, let $f(x)$ be the eigen polynomial or characteristic polynomial of $A_2^{-1} A_1$ and $y = -b_{22} x$. Then,

$$f(-\frac{y}{b_{11}}) = g(y) = -\frac{1}{b_{11}^3} y(y^2 + (a_{11} - \beta)y + (-b_{11}\alpha + b_{22}b_{11})).$$

Since $b_{11} \neq 0$, the equation $g(y) = 0$ and the equation $f(x) = 0$ have three different real roots from Lemma 3.4.

(1-3) $a_{22} = b_{11} = 0$.

T is given by

$$T = \left(\begin{pmatrix} a_{11} & 0 & 1 \\ 0 & 0 & 0 \\ 1 & 0 & 0 \end{pmatrix} ; \begin{pmatrix} 0 & 0 & 0 \\ 0 & b_{22} & 1 \\ 0 & 1 & 0 \end{pmatrix} ; \begin{pmatrix} c_{11} & c_{12} & 0 \\ c_{21} & c_{22} & 0 \\ 0 & 0 & 0 \end{pmatrix} \right).$$

Let $A_1 \leftarrow A_1 + A_2 + \alpha E_{21}$ and $A_2 \leftarrow A_2 + \beta E_{21}$, and let $f(x)$ be the eigen-polynomial of $A_2^{-1}A_1$ and $y = (\alpha - b_{22} - a_{11})x$. Then,

$$f\left(\frac{y}{\alpha - a_{11} - b_{22}}\right) = g(y) = \frac{1}{(\alpha - a_{11} - b_{22})^3}y(y + a_{11} + b_{22} - \alpha)(y + a_{11} - \beta).$$

Thus, the equation $g(y) = 0$ and the equation $f(x) = 0$ have three real roots for appropriate α and β. Thus, $A_1^{-1}A_2$ is diagonalizable, and by Lemma 3.2, $\mathrm{rank}(T) \leq 5$.

(2) The case of $M \in G_2$

As stated in Fact 3.1, by $RC(1 \leftrightarrow 2)$, the members of Group 2 become members of G_2. Furthermore, A_1 and A_2 exchange mutually. This fact implies that the proof for G_1 is applicable to G_2.

(3) The case of $M \in G_3$
Let

$$T = (A_1; A_2; A_3) = \left(\begin{pmatrix} a & 0 & 1 \\ 0 & b & 0 \\ 1 & 0 & 0 \end{pmatrix}; \begin{pmatrix} c & 0 & 0 \\ 0 & d & 0 \\ 0 & 1 & 0 \end{pmatrix}; \begin{pmatrix} s & 0 & 0 \\ 0 & w & 0 \\ 0 & 0 & 0 \end{pmatrix}\right),$$

$$T_1 = \left(\begin{pmatrix} a-1 & 0 & 0 \\ 0 & b & 0 \\ 0 & 0 & 0 \end{pmatrix}; \begin{pmatrix} c & 0 & 0 \\ 0 & d-1 & 0 \\ 0 & 0 & 0 \end{pmatrix}; \begin{pmatrix} s & 0 & 0 \\ 0 & w & 0 \\ 0 & 0 & 0 \end{pmatrix}\right),$$

$$T_2 = \left(\begin{pmatrix} 1 & 0 & 1 \\ 0 & 0 & 0 \\ 1 & 0 & 1 \end{pmatrix}; \begin{pmatrix} 0 & 0 & 0 \\ 0 & 1 & 1 \\ 0 & 1 & 1 \end{pmatrix}; \begin{pmatrix} 0 & 0 & 0 \\ 0 & 0 & 0 \\ 0 & 0 & 0 \end{pmatrix}\right),$$

and

$$T_3 = \left(\begin{pmatrix} 0 & 0 & 0 \\ 0 & 0 & 0 \\ 0 & 0 & -1 \end{pmatrix}; \begin{pmatrix} 0 & 0 & 0 \\ 0 & 0 & 0 \\ 0 & 0 & -1 \end{pmatrix}; \begin{pmatrix} 0 & 0 & 0 \\ 0 & 0 & 0 \\ 0 & 0 & 0 \end{pmatrix}\right).$$

Then, $T = T_1 + T_2 + T_3$ and $\mathrm{rank}(T) \leq \mathrm{rank}(T_1) + \mathrm{rank}(T_1) + \mathrm{rank}(T_1) = 2 + 2 + 1 = 5$.

This completes the proof. Next, we prove the existence of tensors with rank 5.

Proposition 3.8 *For the following tensor* $\mathrm{rank}(T) = 5$.

$$T = \left(\begin{pmatrix} 0 & 1 & 0 \\ 0 & 0 & 0 \\ 0 & 0 & 0 \end{pmatrix}; \begin{pmatrix} 0 & 0 & 0 \\ 0 & 0 & 1 \\ 0 & 0 & 0 \end{pmatrix}; \begin{pmatrix} 1 & 0 & 0 \\ 0 & 1 & 0 \\ 0 & 0 & 1 \end{pmatrix}\right).$$

Proof If $\text{rank}(T) \leq 4$, there are rank-1 tensors D_1, D_2, D_3, D_4 such that $\langle A, B, C \rangle \subset \langle D_1, D_2, D_3, D_4 \rangle$. Since A and B are rank-1 matrices, $\langle D_1, D_2, D_3, D_4 \rangle = \langle A, B, D_s, D_t \rangle$. Hence, $C = xA + yB + zD_s + wD_t$ for some real numbers x, y, z and w. Then, $\text{rank}(C - xA - yB) = 3$ and $\text{rank}(zD_s + wD_t) \leq 2$, which is a contradiction.

From Propositions 3.7 and 3.8, we have proved Proposition 3.6.

Chapter 4
Absolutely Nonsingular Tensors and Determinantal Polynomials

In this chapter, we define absolute nonsingularity for 3-tensors over $\mathbb{R}$ of format $n \times n \times m$ and state a criterion for the existence of an $n \times n \times m$ absolutely nonsingular tensor in terms of Hurwitz–Radon numbers. The zero locus of the determinantal polynomial defined by an $n \times n \times m$ tensor is also discussed.

4.1 Absolutely Nonsingular Tensors

Let $T = (T_1; \ldots; T_m)$ be an $n \times n \times m$-tensor over $\mathbb{R}$. It sometimes happens that there is no singular matrix in the subspace of the vector space of $n \times n$-matrices over $\mathbb{R}$ spanned by $T_1, \ldots, T_m$ except the zero matrix. In view of this fact, we provide the following definition.

Definition 4.1 Let $T = (T_1; \ldots; T_m)$ be an $n \times n \times m$-tensor over $\mathbb{R}$. If

$$\sum_{i=1}^{m} a_i T_i$$

is nonsingular for any $(a_1, \ldots, a_m) \in \mathbb{R}^m \setminus \{0\}$, we say that T is absolutely nonsingular.

Example 4.1 (1) $\left(\begin{pmatrix} 1 & 0 \\ 0 & 1 \end{pmatrix} ; \begin{pmatrix} 0 & -1 \\ 1 & 0 \end{pmatrix} \right)$ is an absolutely nonsingular $2 \times 2 \times 2$-tensor.

(2)

$$\left(E_4; \begin{pmatrix} 0 & -1 & 0 & 0 \\ 1 & 0 & 0 & 0 \\ 0 & 0 & 0 & -1 \\ 0 & 0 & 1 & 0 \end{pmatrix} ; \begin{pmatrix} 0 & 0 & -1 & 0 \\ 0 & 0 & 0 & 1 \\ 1 & 0 & 0 & 0 \\ 0 & -1 & 0 & 0 \end{pmatrix} ; \begin{pmatrix} 0 & 0 & 0 & 1 \\ 0 & 0 & 1 & 0 \\ 0 & -1 & 0 & 0 \\ -1 & 0 & 0 & 0 \end{pmatrix} \right)$$

is an absolutely nonsingular $4 \times 4 \times 4$-tensor.

© The Author(s) 2016
T. Sakata et al., *Algebraic and Computational Aspects of Real Tensor Ranks*,
JSS Research Series in Statistics, DOI 10.1007/978-4-431-55459-2_4

Note that (1) corresponds to the complex numbers and (2) corresponds to the quaternions.

Remark 4.1 Let $T = (T_1; \ldots; T_m)$ be an $n \times n \times m$-tensor with $m \geq 2$. If T_1 is singular, then T is not absolutely nonsingular for obvious reason. If T_1 is nonsingular and λ is an eigenvalue of $T_1^{-1}T_2$, then $\lambda T_1 - T_2$ is singular. Therefore T is not absolutely nonsingular.

In particular, if n is odd and $m \geq 2$, then there is no $n \times n \times m$ absolutely nonsingular tensor.

Next we state a criterion for a tensor to be absolutely nonsingular.

Proposition 4.1 *Let $T = (T_1; \ldots; T_m)$ be an $n \times n \times m$-tensor over $\mathbb{R}$ with $m \geq 2$. Then the following conditions are equivalent, where $S^d := \{(a_1, \ldots, a_{d+1}) \mid a_i \in \mathbb{R}, \sum_{i_1}^{d+1} a_i^2 = 1\}$.*

(1) T is absolutely nonsingular.
(2) For any $a = (a_1, \ldots, a_m) \in S^{m-1}$, $\sum_{k=1}^{m} a_k T_k$ is nonsingular.
(3) $\min_{a=(a_1,\ldots,a_m)\in S^{m-1}} |\det(\sum_{k=1}^{m} a_k T_k)| > 0$.
(4) For any $a = (a_1, \ldots, a_m) \in S^{m-1}$ and for any $b \in S^{n-1}$, $(\sum_{k=1}^{m} a_k T_k) b \neq 0$.
(5) For any $b \in S^{n-1}$, $T_1 b, \ldots, T_m b$ are linearly independent.

Moreover, if $T_1 = E_n$, then the above conditions are equivalent to the following condition.

(6) For any $b \in S^{n-1}$, the orthogonal projections of $T_2 b, \ldots, T_m b$ to $\langle b \rangle^{\perp}$ are linearly independent.

Proof (1) $\Longleftrightarrow$ (2) $\Longleftrightarrow$ (4) $\Longleftrightarrow$ (5) and (3)$\Rightarrow$(2) are straightforward. If the condition (2) is valid, then $\det(\sum_{k=1}^{m} a_k T_k) \neq 0$ for any $a = (a_1, \ldots, a_m) \in S^{m-1}$. Since S^{m-1} is compact in the Euclidean topology and

$$f : S^{m-1} \to \mathbb{R}, \quad a = (a_1, \ldots, a_m) \mapsto |\det\left(\sum_{k=1}^{m} a_k T_k\right)|$$

is a continuous map, the minimum value of this map exists. Since $f(a) > 0$ for any $a \in S^{m-1}$ by (2), $\min_{a=(a_1,\ldots,a_m)\in S^{m-1}} f(a) > 0$.

Now assume that $T_1 = E_n$. In general, $a_1, \ldots, a_r \in \mathbb{R}^n$ are linearly independent if and only if $a_1 \neq 0$ and the orthogonal projections of $a_2, \ldots, a_r$ to $\langle a_1 \rangle^{\perp}$ are linearly independent. Since $T_1 b = b$, (5) $\Longleftrightarrow$ (6) follows.

By using the characterization (3), we show that the set of absolutely nonsingular tensors is an open subset of the set of $n \times n \times m$-tensors over $\mathbb{R}$, $\mathbb{R}^{n \times n \times m}$. First, we note the following fundamental fact.

Lemma 4.1 *Let X and Y be topological spaces and let $f : X \times Y \to \mathbb{R}$ be a continuous map. Suppose that X is compact. Then $g : Y \to \mathbb{R}$, $g(y) = \min_{x \in X} f(x, y)$ is continuous.*

Proposition 4.2 *The set of $n \times n \times m$ absolutely nonsingular tensors over $\mathbb{R}$ is an open subset of $\mathbb{R}^{n \times n \times m}$.*

Proof Consider $f \colon S^{m-1} \times \mathbb{R}^{n \times n \times m} \to \mathbb{R}$, $f((a_1, \ldots, a_m), (T_1; \ldots; T_m)) = |\det (\sum_{k=1}^{m} a_k T_k)|$. Since f is continuous, we see, by Lemma 4.1, that $g \colon \mathbb{R}^{n \times n \times m} \to \mathbb{R}$, $g(T) = \min_{a \in S^{m-1}} f(a, T)$ is a continuous map. Since the set of $n \times n \times m$ absolutely nonsingular tensors is the inverse image of $\{x \in \mathbb{R} \mid x > 0\}$, with respect to g, by (3) of Proposition 4.1, we see that the set of $n \times n \times m$ absolutely nonsingular tensors is an open subset of $\mathbb{R}^{n \times n \times m}$.

Consider the case where $m = n$ and let $T = (T_1; \ldots; T_n)$ be an $n \times n \times n$-tensor. By (1) $\iff$ (4) of Proposition 4.1, we see that T is absolutely nonsingular if and only if for any $a = (a_1, \ldots, a_n) \in S^{n-1}$, $b = (b_1, \ldots, b_n) \in S^{n-1}$, $(\sum_{k=1}^{n} a_k T_k) b \neq 0$. Now let $T' = (T'_1; \ldots; T'_n)$ be the $n \times n \times n$ tensor obtained by rotating T by 90° with the axis parallel to the columns. Then,

$$\left(\sum_{k=1}^{n} a_k T_k \right) b = \left(\sum_{k=1}^{n} b_k T'_{n-k+1} \right) a.$$

Therefore, T is absolutely nonsingular if and only if T' is absolutely nonsingular, i.e., absolute nonsingularity does not depend on the direction from which one looks at a cubic tensor.

4.2 Hurwitz–Radon Numbers and the Existence of Absolutely Nonsingular Tensors

In the previous section, we defined absolutely nonsingular tensors and noted that there is no $n \times n \times m$ absolutely nonsingular tensor if n is odd and $m \geq 2$. In this section, we state a criterion for the existence of absolutely nonsingular tensors of size $n \times n \times m$ in terms of n and m.

First we define the Hurwitz–Radon family.

Definition 4.2 Let $\{A_1, \ldots, A_s\}$ be a family of $n \times n$ matrices with entries in $\mathbb{R}$. If

(1) $A_i A_i^\top = E_n$ for $1 \leq i \leq s$,
(2) $A_i = -A_i^\top$ for $1 \leq i \leq s$ and
(3) $A_i A_j = -A_j A_i$ for $i \neq j$,

then we say that $\{A_1, \ldots, A_s\}$ is a Hurwitz–Radon family of order n.

The following results immediately follow from the definition.

Lemma 4.2 *A subfamily of a Hurwitz–Radon family is a Hurwitz–Radon family.*

Lemma 4.3 *If* $\{A_1, \ldots, A_s\}$ *is a Hurwitz–Radon family of order n, then* $\{E_t \otimes_{kr} A_1, \ldots, E_t \otimes_{kr} A_s\}$ *is a Hurwitz–Radon family of order nt, where* $\otimes_{kr}$ *denote the Kronecker product.*

Next we note the following lemma which is easily verified.

Lemma 4.4 *Let* $\{A_1, \ldots, A_s\}$ *be a Hurwitz–Radon family of order n. Set* $A_{s+1} = E_n$. *Then for any* $a_1, \ldots, a_{s+1} \in \mathbb{R}$,

$$\left(\sum_{k=1}^{s+1} a_k A_k\right)\left(\sum_{k=1}^{s+1} a_k A_k\right)^{\top} = (a_1^2 + \cdots + a_s^2 + a_{s+1}^2)E_n.$$

In particular, $(A_1; \ldots; A_s; E_n)$ *is an* $n \times n \times (s+1)$ *absolutely nonsingular tensor and therefore, if there exists a Hurwitz–Radon family of order n with s members, then there exists an* $n \times n \times (s+1)$ *absolutely nonsingular tensor.*

Next we state the following:

Definition 4.3 Let n be a positive integer. Set $n = (2a + 1)2^{b+4c}$, where a, b, and c are integers with $0 \le b < 4$. Then we define $\rho(n) := 8c + 2^b$.

$\rho(n)$ is called the Hurwitz–Radon number and ρ is called the Hurwitz–Radon function.

The following fact directly follows from the definition.

Lemma 4.5 $\rho(2^s) = 2^s$ *for* $s = 0, 1, 2, 3$ *and* $\rho(2^{4t}) = \rho(2^{4t-1}) + 1$, $\rho(2^{4t+1}) = \rho(2^{4t-1}) + 2$, $\rho(2^{4t+2}) = \rho(2^{4t-1}) + 4$ *and* $\rho(2^{4t+3}) = \rho(2^{4t-1}) + 8$ *for any positive integer t.*

Next we state the existence of the Hurwitz–Radon family of order n with member $\rho(n) - 1$. Set

$$A = \begin{pmatrix} 0 & 1 \\ -1 & 0 \end{pmatrix}, \quad P = \begin{pmatrix} 0 & 1 \\ 1 & 0 \end{pmatrix}, \quad Q = \begin{pmatrix} 1 & 0 \\ 0 & -1 \end{pmatrix}.$$

Then one can verify the following results by routine calculations.

Lemma 4.6 (Geramita and Seberry 1979, Proposition 1.5)

(1) $\{A\}$ *is a Hurwitz–Radon family of order 2.*

(2) $\{A \otimes_{kr} E_2, P \otimes_{kr} A, Q \otimes_{kr} A\}$ *is a Hurwitz–Radon family of order 4.*

(3) $\{E_2 \otimes_{kr} A \otimes_{kr} E_2, E_2 \otimes_{kr} P \otimes_{kr} A, Q \otimes_{kr} Q \otimes_{kr} A, P \otimes_{kr} Q \otimes_{kr} A, A \otimes_{kr} P \otimes_{kr} Q, A \otimes_{kr} P \otimes_{kr} P, A \otimes_{kr} Q \otimes_{kr} E_2\}$ *is a Hurwitz–Radon family of order 8.*

Lemma 4.7 (Geramita and Seberry 1979, Theorem 1.6) *Let* $\{M_1, \ldots, M_s\}$ *be a Hurwitz–Radon family of order n. Then*

(1) $\{A \otimes_{kr} E_n, Q \otimes_{kr} M_1, \ldots, Q \otimes_{kr} M_s\}$ *is a Hurwitz–Radon family of order* $2n$.

(2) *Moreover, if* $\{L_1, \ldots, L_t\}$ *is a Hurwitz–Radon family of order* m, *then* $\{P \otimes_{kr} M_1 \otimes_{kr} E_m, \ldots, P \otimes_{kr} M_s \otimes_{kr} E_m, Q \otimes_{kr} E_n \otimes_{kr} L_1, \ldots, Q \otimes_{kr} E_n \otimes_{kr} L_t, A \otimes_{kr} E_{mn}\}$ *is a Hurwitz–Radon family of order* $2mn$.

Theorem 4.1 *Let* n *be a positive integer. Then there is a Hurwitz–Radon family of order* n *with* $\rho(n) - 1$ *members.*

Proof If $n = 1$, there is nothing to prove. Therefore we assume that $n > 1$. By Lemma 4.3, we may assume that n is a power of 2. Set $n = 2^s$. We obtain the proof by induction on s. The cases where $s = 1, 2, 3$ are covered by Lemmas 4.5 and 4.6.

Assume that $s \geq 4$ and set $s = 4c + b$ with $0 \leq b < 4$. By induction hypothesis, we see that there is a Hurwitz–Radon family of order 2^{4c-1} with $\rho(2^{4c-1}) - 1$ members. If $s = 4c$, then we see by Lemma 4.7 (1) and Lemma 4.5 that there is a Hurwitz–Radon family of order 2^s with $\rho(2^s) - 1$ members. If $s = 4c + 1$ ($s = 4c + 2, 4c + 3$ resp.), then we see by Lemma 4.6 (1), Lemma 4.7 (2), and Lemma 4.5 (Lemma 4.6 (2), Lemma 4.7 (2) and Lemma 4.5, Lemma 4.6 (3), Lemma 4.7 (2) and Lemma 4.5, resp.) that there is a Hurwitz–Radon family of order 2^s with $\rho(2^s) - 1$ members.

By this Theorem and Lemma 4.4, we see the following fact.

Corollary 4.1 *If* $m \leq \rho(n)$ *then there exists an* $n \times n \times m$ *absolutely nonsingular tensor.*

Next we prove the converse of the above result. First we cite the following result.

Theorem 4.2 (Adams 1962) *There do not exist* $\rho(n)$ *linearly independent vector fields on* S^{n-1}.

By Proposition 4.1 (6), if there exists an $n \times n \times m$ absolutely nonsingular tensor, then there exist $m - 1$ linearly independent vector fields on S^{n-1}. Therefore by Theorem 4.2, we see the following fact.

Corollary 4.2 *If there exists an* $n \times n \times m$ *absolutely nonsingular tensor, then* $m \leq \rho(n)$.

By summing up the above results, we see the following result.

Theorem 4.3 *There exists an* $n \times n \times m$ *absolutely nonsingular tensor if and only if* $m \leq \rho(n)$.

4.3 Bilinear Maps and Absolutely Full Column Rank Tensors

A square matrix is nonsingular if and only if it is a full column rank matrix. Therefore by generalizing the notion of an absolutely nonsingular tensor, we arrive at the following notion.

Definition 4.4 Let $T = (T_1; \ldots; T_m)$ be a $u \times n \times m$ tensor over $\mathbb{R}$. T is called an Absolutely full column rank tensor if

$$\mathrm{rank}\left(\sum_{k=1}^{m} a_k T_k\right) = n$$

for any $(a_1, \ldots, a_m) \in \mathbb{R}^m \setminus \{\mathbf{0}\}$.

We see the following fact in the same way as Proposition 4.1.

Proposition 4.3 *Let u, n, and m be positive integers with $u \geq n$ and let $T = (T_1; \ldots; T_m)$ be a $u \times n \times m$ tensor over $\mathbb{R}$. Then the following conditions are equivalent.*

(1) T is Absolutely full column rank.
(2) For any $\mathbf{a} = (a_1, \ldots, a_m) \in S^{m-1}$, $\sum_{k=1}^{m} a_k T_k$ is full column rank.
(3) $\min_{\mathbf{a} = (a_1, \ldots, a_m) \in S^{m-1}} (\text{the maximum of the absolute values of n-minors of } \sum_{k=1}^{m} a_k T_k) > 0$.
(4) For any $\mathbf{a} = (a_1, \ldots, a_m) \in \mathbb{R}^m \setminus \{\mathbf{0}\}$ and any $\mathbf{b} \in \mathbb{R}^n \setminus \{\mathbf{0}\}$, $\left(\sum_{k=1}^{m} a_k T_k\right)\mathbf{b} \neq \mathbf{0}$.

We can also see the following fact in the same way as Proposition 4.2.

Proposition 4.4 *The set of $u \times n \times m$ Absolutely full column rank tensors over $\mathbb{R}$ is an open subset of $\mathbb{R}^{u \times n \times m}$.*

Note that by defining

$$f_T(\mathbf{x}, \mathbf{y}) = \left(\sum_{k=1}^{m} x_k T_k\right)\mathbf{y}$$

for a $u \times n \times m$ tensor $T = (T_1; \ldots; T_m)$, where $(x_1, \ldots, x_m) = \mathbf{x}$, one defines naturally a one-to-one correspondence between the set of $u \times n \times m$ tensors and the bilinear maps $\mathbb{R}^m \times \mathbb{R}^n \to \mathbb{R}^u$.

Definition 4.5 Let $f : \mathbb{R}^m \times \mathbb{R}^n \to \mathbb{R}^u$ be a bilinear map. We say that f is nonsingular if $f(\mathbf{x}, \mathbf{y}) = \mathbf{0}$ implies that $\mathbf{x} = \mathbf{0}$ or $\mathbf{y} = \mathbf{0}$.

Set

$$m\#n = \min\{u \mid \text{there exists a nonsingular bilinear map } \mathbb{R}^m \times \mathbb{R}^n \to \mathbb{R}^u.\}$$

Then, by Proposition 4.3, we see the following:

Corollary 4.3 *There exists a $u \times n \times m$ Absolutely full column rank tensor if and only if $m\#n \leq u$.*

A complete criterion for the existence of a nonsingular bilinear map is not known. See Shapiro 2000, Chap. 12. We just comment on the following fact.

Set

$$m \circ n = \min \left\{ u \mid \text{if } u - m < k < n, \text{ then the binomial coefficient } \binom{u}{k} \text{ is even.} \right\}$$

Then, the following inequalities are known:

$$\max\{m, n\} \leq m \circ n \leq m\#n \leq m + n - 1$$

Further, it is known that if $\min\{m, n\} \leq 9$, then $m \circ n = m\#n$. See Shapiro 2000, Chap. 12.

4.4 Determinantal Polynomials and Absolutely Nonsingular Tensors

In this section, we consider determinantal polynomials of the form

$$\det \left(\sum_{k=1}^{m} x_k T_k \right),$$

where $T = (T_1; \ldots; T_m)$ is an $n \times n \times m$-tensor and $x_1, \ldots, x_m$ are indeterminates.

Here we recall the real Nullstellensatz. First we recall the following:

Definition 4.6 Let A be a commutative ring and I an ideal of A.

$$\sqrt[R]{I} = \left\{ a \in A \mid \exists k \in \mathbb{N} \; \exists b_1, \ldots, b_t \in A \text{ such that } a^{2k} + b_1^2 + \cdots + b_t^2 \in I \right\}$$

is called the real radical of I.

The real Nullstellensatz is the following (see Sect. 6.2 for the definition of $\mathbb{I}$ and $\mathbb{V}$):

Theorem 4.4 (Bochnak et al. 1998, Theorem 4.1.4, and Corollary 4.1.8) *Let I be an ideal of $\mathbb{R}[x_1, \ldots, x_m]$, the polynomial ring with m variables over $\mathbb{R}$. Then $\mathbb{I}(\mathbb{V}(I)) = \sqrt[R]{I}$.*

Let T be an $n \times n \times m$ tensor over $\mathbb{R}$ and let $f(x_1, \ldots, x_m)$ be the determinantal polynomial defined by T. Then T is absolutely nonsingular if and only if

$$\{(a_1, \ldots, a_m) \in \mathbb{R}^m \mid f(a_1, \ldots, a_m) = 0\} = \{(0, \ldots, 0)\}.$$

Therefore, by the real Nullstellensatz Theorem 4.4, we see the following fact:

Proposition 4.5 *Let T be an $n \times n \times m$-tensor over $\mathbb{R}$ and $f(x_1, \ldots, x_m)$ the determinantal polynomial defined by T. Then the following conditions are equivalent:*

(1) T is absolutely nonsingular.
(2) The real radical $\sqrt[R]{\langle f \rangle}$ of the principal ideal generated by f is $\langle x_1, \ldots, x_m \rangle$.

Next we consider the irreducibility of f.

Definition 4.7 Let $\mathbb{K}$ be a field and let $x_1, \ldots, x_t$ be indeterminates. A polynomial $g \in \mathbb{K}[x_1, \ldots, x_t]$ is called *absolutely prime* if it is irreducible in $\overline{\mathbb{K}}[x_1, \ldots, x_t]$, where $\overline{\mathbb{K}}$ is the algebraic closure of $\mathbb{K}$.

Theorem 4.5 (Heintz and Sieveking 1981) *Let $\mathbb{K}$ be a field with $\mathrm{char}\mathbb{K} = 0$, let $x_1, \ldots, x_t$ be indeterminates where $t \geq 2$, and let d be a positive integer. Then, there is a dense Zariski open subset U of the set of polynomials with degree at most d such that every element of U is absolutely prime.*

Since there is a one-to-one correspondence

$$g(x_1, \ldots, x_t, x_{t+1}) \mapsto g(x_1, \ldots, x_t, 1)$$

between the set of homogeneous polynomials with degree d and $t + 1$ variables and the set of polynomials with degree at most d and t variables, we see the following:

Corollary 4.4 *Let $\mathbb{K}$ be a field with $\mathrm{char}\mathbb{K} = 0$, let $x_1, \ldots, x_t$ be indeterminates with $t \geq 3$, and let d be a positive integer. Then, there is a dense Zariski open subset U of the set of homogeneous polynomials with degree d such that every element of U is irreducible.*

Suppose that $m \geq 3$ and let $P(m, d)$ be the set of degree d homogeneous polynomials with coefficients in $\mathbb{R}$ and variables $x_1, \ldots, x_m$. Since $\mathbb{R}^{n \times n \times m} \to P(m, n)$, $T = (T_1; \ldots; T_m) \mapsto \det\left(\sum_{k=1}^m x_k T_k\right)$ is a polynomial map (see Definition 6.4), we see that the inverse image of U of Corollary 4.4 is a Zariski open subset of $\mathbb{R}^{n \times n \times m}$. In fact, this inverse image is not empty by Sumi et al. 2015a, Theorem 5.1, and Proposition 5.2, and therefore it is dense.

Moreover, we see the following fact.

Theorem 4.6 *Suppose that $3 \leq m \leq n$ and let $x_1, \ldots, x_m$ be indeterminates. Then there are Euclidean open subsets $\mathcal{O}_1$ and $\mathcal{O}_2$ of $\mathbb{R}^{n \times n \times m}$ with the following properties.*

(1) $\mathcal{O}_1 \cup \mathcal{O}_2$ is a dense subset of $\mathbb{R}^{n \times n \times m}$ in the Euclidean topology.
(2) If $T = (T_1; \ldots; T_m) \in \mathcal{O}_1 \cup \mathcal{O}_2$, then $\det\left(\sum_{k=1}^m x_k T_k\right)$ is an irreducible polynomial.
(3) If $T \in \mathcal{O}_1$, then T is absolutely nonsingular, i.e.,

$$\sqrt[R]{\left\langle \det\left(\sum_{k=1}^m x_k T_k\right)\right\rangle} = \langle x_1, \ldots, x_m \rangle.$$

(4) If $T \in \mathcal{O}_2$, then

$$\sqrt[R]{\left\langle \det\left(\sum_{k=1}^{m} x_k T_k\right)\right\rangle} = \left\langle \det\left(\sum_{k=1}^{m} x_k T_k\right)\right\rangle.$$

Proof We use the notation of Sumi et al. 2015a. Set $\mathcal{O}_1 = \{T = (T_1; \ldots; T_m) \in \mathbb{R}^{n \times n \times m} \mid T$ is absolutely nonsingular and $\det\left(\sum_{k=1}^{m} x_k T_k\right)$ is irreducible$\}$ and $\mathcal{O}_2 = \{T = (T_1; \ldots; T_m) \in \mathbb{R}^{n \times n \times m} \mid T_m$ is nonsingular and $(T_1 T_m^{-1}; \ldots; T_{m-1} T_m^{-1}) \in \mathcal{U} \cap \mathcal{C}\}$. Then by the definitions of $\mathcal{U}$ and $\mathcal{C}$, we see that $\mathcal{O}_1 \cup \mathcal{O}_2$ is a Euclidean dense open subset of $\mathbb{R}^{n \times n \times m}$ and $\det\left(\sum_{k=1}^{m} x_k T_k\right)$ is irreducible for any $T = (T_1; \ldots; T_m) \in \mathcal{O}_1 \cup \mathcal{O}_2$.

If $T \in \mathcal{O}_1$, then

$$\sqrt[R]{\left\langle \det\left(\sum_{k=1}^{m} x_k T_k\right)\right\rangle} = \langle x_1, \ldots, x_m \rangle,$$

since T is absolutely nonsingular. Now, suppose that $T = (T_1; \ldots; T_m) \in \mathcal{O}_2$. We have to show that $g(x_1, \ldots, x_m) \in \langle \det\left(\sum_{k=1}^{m} x_k T_k\right)\rangle$ if $g(x_1, \ldots, x_m) \in \mathbb{R}[x_1, \ldots, x_m]$ and $g(a_1, \ldots, a_m) = 0$ for any $(a_1, \ldots, a_m) \in \mathbb{R}^m$ with $\det\left(\sum_{k=1}^{m} a_k T_k\right) = 0$.

Assume the contrary. Then, since $\det\left(\sum_{k=1}^{m} x_k T_k\right)$ is an irreducible polynomial, we see that there are $f_1(x_1, \ldots, x_m)$ and $f_2(x_1, \ldots, x_m) \in \mathbb{R}[x_1, \ldots, x_m]$ and a nonzero polynomial $h(x_1, \ldots, x_{m-1}) \in \mathbb{R}[x_1, \ldots, x_{m-1}]$ such that

$$f_1(x_1, \ldots, x_m) \det\left(\sum_{k=1}^{m} x_k T_k\right) + f_2(x_1, \ldots, x_m) g(x_1, \ldots, x_m) = h(x_1, \ldots, x_{m-1}).$$

$$(4.4.1)$$

Take $a = (a_1, \ldots, a_m) \in \mathbb{R}^m$ with $\det\left(M(a, T T_m^{-1})\right) < 0$. Then there is an open neighborhood U of $(a_1, \ldots, a_{m-1})$ in $\mathbb{R}^{m-1}$ and a mapping $\mu \colon U \to \mathbb{R}$ such that

$$\det\left(M\left(\begin{pmatrix} y \\ \mu(y) \end{pmatrix}, T T_m^{-1}\right)\right) = 0$$

for any $y \in U$. Since $h \neq 0$, we can take $(b_1, \ldots, b_{m-1}) \in U$ such that $h(b_1, \ldots, b_{m-1}) \neq 0$. Set $b_m = \mu(b_1, \ldots, b_{m-1})$. Then, since $\det\left(\sum_{k=1}^{m} b_k T_k\right) = 0$, $g(b_1, \ldots, b_m) = 0$ by the assumption of g. This contradicts to (4.4.1).

Chapter 5
Maximal Ranks

In this chapter, we consider the maximal rank of tensors with format $(m, n, 2)$ or $(m, n, 3)$. We also introduce upper bounds and lower bounds of the ranks of tensors with format (m, n, p), where $m, n, p \geq 3$.

5.1 Classification and Maximal Rank of $m \times n \times 2$ Tensors

The set $\mathbb{K}^{m \times n \times p}$ or $T_\mathbb{K}(m, n, p)$ denotes the set of all tensors with size $m \times n \times p$ over a field $\mathbb{K}$. The triad (m, n, p) is also called the *format* of the set. A tensor with format (m, n, p) implies a tensor with size $m \times n \times p$. The set $T_\mathbb{K}(m, n, p)$ has an action of $GL(m, \mathbb{K}) \times GL(n, \mathbb{K}) \times GL(p, \mathbb{K})$, which preserves rank. Let $\max.\mathrm{rank}_\mathbb{K}(m, n, p)$ denote the maximal rank of tensors of $T_\mathbb{K}(m, n, p)$.

First, we recall the Jordan normal form. Let $\mathbb{K}$ be an algebraically closed field. Let E_n be the $n \times n$ identity matrix and let

$$
J_n = \begin{pmatrix} 0 & 1 & & 0 \\ \vdots & \ddots & \ddots & \\ 0 & \cdots & 0 & 1 \\ 0 & \cdots & 0 & 0 \end{pmatrix}
$$

be an $n \times n$ superdiagonal matrix. For our convenience, we assume that J_1 is the null matrix. An $n \times n$ matrix A whose elements are in $\mathbb{K}$ is similar to a Jordan matrix

$$
\mathrm{Diag}(\lambda_1 E_{n_1} + J_{n_1}, \ldots, \lambda_t E_{n_t} + J_{n_t}),
$$

where $\lambda_i \in \mathbb{K}$ and $n_i \geq 1$, i.e., there exists $P \in GL(n, \mathbb{K})$ such that $P^{-1}AP$ is equal to the above Jordan matrix. A diagonal element λ_i is an eigenvalue of A. In the case where $\mathbb{K} = \mathbb{C}$, if $A^*A = AA^*$, where A^* is the complex conjugate transpose, then A

© The Author(s) 2016
T. Sakata et al., *Algebraic and Computational Aspects of Real Tensor Ranks*,
JSS Research Series in Statistics, DOI 10.1007/978-4-431-55459-2_5

is said to be *normal*. If A is normal, then A is diagonalizable over $\mathbb{C}$, i.e., A is similar to a diagonal matrix.

Let A be an $n \times n$ matrix whose elements are in $\mathbb{R}$ such that the characteristic polynomial $\det(\lambda E_n - A)$ of A is

$$\prod_{i=1}^{s}(\lambda - \mu_i) \prod_{i=1}^{t}(\lambda^2 - 2a_i\lambda + a_i^2 + b_i^2),$$

where $\mu_i, a_i, b_i \in \mathbb{R}$. Put

$$C_m(c, s) = E_m \otimes \begin{pmatrix} c & -s \\ s & c \end{pmatrix} = \mathrm{Diag}\left(\begin{pmatrix} c & -s \\ s & c \end{pmatrix}, \ldots, \begin{pmatrix} c & -s \\ s & c \end{pmatrix}\right),$$

which is a $2m \times 2m$ square matrix. If $(a + b\sqrt{-1})E_m + J_m$ is a Jordan block of A over $\mathbb{C}$, where $a, b \in \mathbb{R}$, $b \neq 0$, then $(a - b\sqrt{-1})E_m + J_m$ is also a Jordan block of A over $\mathbb{C}$ and $C_m(a, b) + J_m \otimes E_2$ is a Jordan block of A over $\mathbb{R}$. In particular, A is similar to a diagonal matrix

$$\mathrm{Diag}(\mu_1, \ldots, \mu_s, a_1 + b_1\sqrt{-1}, a_1 - b_1\sqrt{-1}, \ldots, a_t + b_t\sqrt{-1}, a_t - b_t\sqrt{-1})$$

over $\mathbb{C}$ if and only if A is similar to

$$\mathrm{Diag}(\mu_1, \ldots, \mu_s, C_1(a_1, b_1), \ldots, C_1(a_t, b_t))$$

over $\mathbb{R}$.

Kronecker and Weierstrass showed that every pair of matrices can be transformed into a canonical pair by pre-multiplication and post-multiplication. In terms of tensors, any 3-tensor $(A; B)$ with format $(m, n, 2)$ is $\mathrm{GL}(m, \mathbb{K}) \times \mathrm{GL}(n, \mathbb{K})$-equivalent to some tensor. A tensor $(A; B)$ with 2 slices has one-to-one correspondence with a homogeneous pencil $\lambda A + \mu B$, where λ and μ are indeterminates.

For tensors $X_k = (A_k; B_k)$ of format $(m_k, n_k, 2)$, $1 \leq k \leq t$,

$$\mathrm{Diag}(X_1, X_2, \ldots, X_t)$$

denotes the tensor

$$\left(\begin{pmatrix} A_1 & & \\ & A_2 & \mathbf{O} \\ & \mathbf{O} & \ddots \\ & & & A_t \end{pmatrix}; \begin{pmatrix} B_1 & & \\ & B_2 & \mathbf{O} \\ & \mathbf{O} & \ddots \\ & & & B_t \end{pmatrix}\right)$$

of format $(m_1 + m_2 + \cdots + m_t, n_1 + n_2 + \cdots + n_t, 2)$. This notation depends on the direction of slices.

Theorem 5.1 (Gantmacher 1959, (30) in Sect. 4, XII) *Let $\mathbb{K}$ be an algebraically closed field. A 3-tensor $(A; B) \in T_{\mathbb{K}}(m, n, 2)$ is $\mathrm{GL}(m, \mathbb{K}) \times \mathrm{GL}(n, \mathbb{K})$-equivalent to a tensor of block diagonal form*

$$\mathrm{Diag}((S_1; T_1), \ldots, (S_r; T_r)),$$

where each $(S_j; T_j)$ is one of the following:

(A) zero tensor $(O; O) \in T_{\mathbb{K}}(k, l, 2)$, $k, l \geq 0$, $(k, l) \neq (0, 0)$,
(B) $(aE_k + J_k; E_k) \in T_{\mathbb{K}}(k, k, 2)$, $k \geq 1$,
(C) $(E_k; J_k) \in T_{\mathbb{K}}(k, k, 2)$, $k \geq 1$,
(D) $((O, E_k); (E_k, O)) \in T_{\mathbb{K}}(k, k + 1, 2)$, $k \geq 1$,
(E) $\left(\begin{pmatrix} O \\ E_k \end{pmatrix}; \begin{pmatrix} E_k \\ O \end{pmatrix} \right) \in T_{\mathbb{K}}(k + 1, k, 2)$, $k \geq 1$.

Moreover, $(A; B) \in T_{\mathbb{R}}(m, n, 2)$ is $\mathrm{GL}(m, \mathbb{R}) \times \mathrm{GL}(n, \mathbb{R})$-equivalent to a tensor of block diagonal form $\mathrm{Diag}((S_1; T_1), \ldots, (S_r; T_r))$, where each $(S_j; T_j)$ is one of (A)–(E) and

(F) $(C_k(c, s) + J_k \otimes E_2; E_{2k}) \in T_{\mathbb{R}}(2k, 2k, 2)$, $s \neq 0$, $k \geq 1$.

This decomposition is called the *Kronecker–Weierstrass canonical form*. It is unique up to permutations of blocks. Each block is called a Kronecker–Weierstrass block. Note that tensors of type (A) include those when $k > 0$ and $l = 0$, or $k = 0$ and $l > 0$, where the direct sum of a tensor with format $(0, l, 2)$ of type (A) and a tensor $(X; Y)$ with format $(s, t, 2)$ implies a tensor $((O, X); (O, Y))$ with format $(s, l + t, 2)$.

For a tensor $(A; B) \in T_{\mathbb{K}}(n, n, 2)$, if the vector space $\langle A, B \rangle$ spanned by A and B contains a nonsingular matrix, then the Kronecker–Weierstrass canonical form does not contain a block of type (A), (D), or (E). We remark that $(aE_k + J_k; E_k)$ and (E_k, J_k) are $\mathrm{GL}(k, \mathbb{K})^{\times 2} \times \mathrm{GL}(2, \mathbb{K})$-equivalent to $(J_k; E_k)$, and that $(C_k(c, s) + J_k \otimes E_2; E_{2k})$ with $s \neq 0$ is $\mathrm{GL}(k, \mathbb{R})^{\times 2} \times \mathrm{GL}(2, \mathbb{R})$-equivalent to $(C_k(0, 1) + J_k \otimes E_2; E_{2k})$.

Example 5.1 A tensor $(A; B) \in T_{\mathbb{R}}(3, 3, 2)$ such that $\langle A, B \rangle$ has no nonsingular matrix is $\mathrm{GL}(3, \mathbb{R})^{\times 2} \times \mathrm{GL}(2, \mathbb{R})$-equivalent to one of the tensors $(A'; B')$ such that $xA' + yB'$ is given by O, $\begin{pmatrix} y\,0\,0 \\ 0\,0\,0 \\ 0\,0\,0 \end{pmatrix}$, $\begin{pmatrix} y & 0 & 0 \\ 0 & ax+y & 0 \\ 0 & 0 & 0 \end{pmatrix}$, $\begin{pmatrix} y\,x\,0 \\ 0\,0\,0 \\ 0\,0\,0 \end{pmatrix}$, $\begin{pmatrix} y\,0\,0 \\ x\,0\,0 \\ 0\,0\,0 \end{pmatrix}$, $\begin{pmatrix} y\,x\,0 \\ 0\,y\,0 \\ 0\,0\,0 \end{pmatrix}$, $\begin{pmatrix} y & -x & 0 \\ x & y & 0 \\ 0 & 0 & 0 \end{pmatrix}$, and $\begin{pmatrix} y\,x\,0 \\ 0\,0\,y \\ 0\,0\,x \end{pmatrix}$. The tensor $(A'; B')$ has rank 0, 1, 2, 2, 2, 3, 3, and 4, respectively.

The rank of a Kronecker–Weierstrass block is given as follows:

Proposition 5.1 (Ja'Ja' 1979; Sumi et al. 2009)

(1) $\mathrm{rank}_{\mathbb{F}}(O; O) = 0$ *and* $\mathrm{rank}_{\mathbb{F}}(a; 1) = 1$.

(2) $\mathrm{rank}_{\mathbb{F}}(aE_k + J_k; E_k) = \mathrm{rank}_{\mathbb{F}}(E_k; J_k) = k + 1$ *for* $k \geq 2$.

(3) $\mathrm{rank}_{\mathbb{F}}((O, E_k); (E_k, O)) = \mathrm{rank}_{\mathbb{F}}\left(\begin{pmatrix} O \\ E_k \end{pmatrix}; \begin{pmatrix} E_k \\ O \end{pmatrix} \right) = k + 1$ *for* $k \geq 1$.

(4) $\mathrm{rank}_{\mathbb{R}}(C_k(c, s) + J_k \otimes E_2; E_{2k}) = 2k + 1$ *if* $s \neq 0$ *and* $k \geq 1$.

Theorem 5.2 (Ja'Ja' 1979) *Let A be a tensor with format* $(m, n, 2)$ *and let B be a tensor of type (D) or (E). Then,* $\mathrm{rank}_{\mathbb{F}}(\mathrm{Diag}(A, B)) = \mathrm{rank}_{\mathbb{F}}(A) + \mathrm{rank}_{\mathbb{F}}(B)$.

In general, the rank of a tensor is not the sum of ranks of its Kronecker–Weierstrass blocks.

Theorem 5.3 (Sumi et al. 2009, Theorem 4.6) *Let A be an* $n \times n$ *matrix and let* $\alpha_{\mathbb{F}}(A, \lambda)$ *be the number of Kronecker–Weierstrass blocks whose sizes are greater than or equal to 2 for an eigenvalue* λ *of A. Then,*

$$\mathrm{rank}_{\mathbb{F}}(E_n; A) = n + \max_{\lambda} \alpha_{\mathbb{F}}(A, \lambda),$$

where we treat $C_k(c, s) + J_k \otimes E_2$ *as a Kronecker–Weierstrass block of size* $2k$ *if* $\mathbb{F} = \mathbb{R}$.

Let A and B be $m \times n$ rectangular matrices. We describe the rank of a tensor $(A; B)$ with its Kronecker–Weierstrass canonical form. Suppose that the Kronecker–Weierstrass canonical form $(S; T)$ of $(A; B)$ has a zero tensor $(O; O)$ of type (A) with format $(m_A, n_A, 2)$, l_D tensors of type (D), and l_E tensors of type (E). Set the part of types (B) and (F) of $(S; T)$ as $(S_B; E)$ and $(S_F; E)$, respectively, and the part of type (C) of $(S; T)$ as $(E; S_C)$. Put

$$\alpha = \max\{\max_{\lambda} \alpha_{\mathbb{C}}(S_B, \lambda), \max_{\lambda} \alpha_{\mathbb{C}}(S_C, \lambda)\}$$

if $\mathbb{F} = \mathbb{C}$ and

$$\alpha = \max\{\max_{\lambda} \alpha_{\mathbb{R}}(S_B, \lambda), \max_{\lambda} \alpha_{\mathbb{R}}(S_C, \lambda), \max_{\lambda} \alpha_{\mathbb{R}}(S_F, \lambda)\}$$

if $\mathbb{F} = \mathbb{R}$, where $\alpha_{\mathbb{F}}$ is as in Theorem 5.3.

Theorem 5.4 (Sumi et al. 2009, Theorem 1.5)

$$\mathrm{rank}_{\mathbb{F}}(A; B) = m - m_A + \alpha + l_D = n - n_A + \alpha + l_E.$$

Similarly, we have the maximal rank.

Theorem 5.5 (Ja'Ja' 1979; Sumi et al. 2009, Theorem 4.3) *If the cardinality of $\mathbb{K}$ is greater than or equal to $\min\{m, n\}$, then*

$$\text{max.rank}_{\mathbb{K}}(m, n, 2) = \min\{n + \left\lfloor \frac{m}{2} \right\rfloor, m + \left\lfloor \frac{n}{2} \right\rfloor, 2m, 2n\}.$$

For a tensor A of format $(a, b, 2)$ and a positive integer n, $A^{\oplus n}$ denotes the tensor

$$\text{Diag}(\overbrace{A, A, \ldots, A}^{n})$$

of format $(na, nb, 2)$.

Moreover, for integers m and n with $m \le n \le 2m$, a tensor of $T_{\mathbb{K}}(m, n, 2)$ having the maximal rank $m + \lfloor n/2 \rfloor$ is $\text{GL}(m, \mathbb{K}) \times \text{GL}(n, \mathbb{K})$-equivalent to

$$\text{Diag}(Y^{\oplus \alpha}, ((0, 1); (1, 0))^{\oplus \beta})$$

if n is even and

- $\text{Diag}(Y^{\oplus \alpha}, ((0, 1); (1, 0))^{\oplus(\beta-1)}, O)$, $O \in T_{\mathbb{K}}(1, 0, 2)$,
- $\text{Diag}(Y^{\oplus(\alpha-1)}, ((0, 1); (1, 0))^{\oplus \beta}, (\mu; 1))$,
- $\text{Diag}(Y^{\oplus(\alpha-1)}, ((0, 1); (1, 0))^{\oplus \beta}, (1; 0))$,
- $\text{Diag}(Y^{\oplus(\alpha-1)}, ((0, 1); (1, 0))^{\oplus(\beta-1)}, ((O, E_2); (E_2, O)))$,
- $\text{Diag}(Y^{\oplus(\alpha-2)}, \left((0, 1); (1, 0))^{\oplus(\beta+1)}, \left(\begin{pmatrix} 0 \\ 1 \end{pmatrix}; \begin{pmatrix} 1 \\ 0 \end{pmatrix}\right)\right)$,
- $\text{Diag}((\lambda E_2 + J_2; E_2)^{\oplus(\alpha-2)}, (\lambda E_3 + J_3; E_3), ((0, 1); (1, 0))^{\oplus \beta})$, or
- $\text{Diag}((E_2; J_2)^{\oplus(\alpha-2)}, (E_3; J_3), ((0, 1); (1, 0))^{\oplus \beta})$

if n is odd, where $\alpha = m - \lfloor n/2 \rfloor$, $\beta = n - m$, and Y is $(\lambda E_2 + J_2; E_2)$ or $(E_2; J_2)$ when $\mathbb{K}$ is algebraically closed, and $(\lambda E_2 + J_2; E_2)$, $(E_2; J_2)$, or $(C_1(c, s); E_2)$ when $\mathbb{K} = \mathbb{R}$.

An $m \times n$ rectangular matrix A of form (D, O) or $(D, O)^{\top}$ depending on the size of m and n, where D is a diagonal matrix, is called a *rectangular diagonal matrix*. For $m \times n$ matrices $A_1, \ldots, A_k$, we say that they are *simultaneously rectangular diagonalizable*, if there exist a nonsingular $m \times m$ matrix P and a nonsingular $n \times n$ matrix Q such that PA_tQ, $1 \le t \le k$, are all rectangular diagonal matrices.

Theorem 5.6 *Let $1 \le f_1 \le f_2 \le f_3$ and $f = (f_1, f_2, f_3)$. Let T and A be 3-tensors with format f and put $T + A = (B_1; \ldots; B_{f_3})$. If $f_1 \times f_2$ matrices $B_1, B_2, \ldots, B_{f_3}$ are simultaneously rectangular diagonalizable, then*

$$\text{rank}_{\mathbb{K}}(T) \le \min\{f_1, f_2\} + \text{rank}_{\mathbb{K}}(A).$$

Proof If $B_1, B_2, \ldots, B_{f_3}$ are simultaneously rectangular diagonalizable, then we see that $\text{rank}_{\mathbb{K}}(T + A) \le \min\{f_1, f_2\}$. Therefore,

$$\text{rank}_{\mathbb{K}}(T) \le \text{rank}_{\mathbb{K}}(T + A) + \text{rank}_{\mathbb{K}}(A) \le \min\{f_1, f_2\} + \text{rank}_{\mathbb{K}}(A).$$

$$\text{Let } Q_1 = E_4,\ Q_2 = \begin{pmatrix} & -1 & & \\ 1 & & & \\ & & -1 & \\ & & & 1 \end{pmatrix},\ Q_3 = \begin{pmatrix} & & -1 & \\ & & & 1 \\ 1 & & & \\ & -1 & & \end{pmatrix},\ Q_4 = \begin{pmatrix} & & & 1 \\ & & 1 & \\ & -1 & & \\ -1 & & & \end{pmatrix},$$

and let $W = \langle Q_1, Q_2, Q_3, Q_4 \rangle$ be a real vector space. The vector space W is closed under the multiplications of matrices and a skew field isomorphic to the quaternions.

Proposition 5.2 *Let A and B be matrices in W. Then, $\mathrm{rank}_{\mathbb{R}}(A; B)$ is equal to 0, 4, or 6. If $\mathrm{rank}_{\mathbb{R}}(A; B) = 0$, then $A = B = O$, and if $\mathrm{rank}_{\mathbb{R}}(A; B) = 6$, then $(A; B)$ is $\mathrm{GL}(4, \mathbb{R})^{\times 2} \times \mathrm{GL}(2, \mathbb{R})$-equivalent to $(Q_1; Q_2)$. Moreover, $\mathrm{rank}_{\mathbb{R}}(A; B) = 4$ if and only if A and B are linearly dependent and $(A, B) \neq (O, O)$.*

Proof Since a singular matrix of W is only zero, $\mathrm{rank}(A) \leq 3$ implies that $A = O$. Since $\mathrm{rank}_{\mathbb{R}}(A; B) \geq \max\{\mathrm{rank}(A), \mathrm{rank}(B)\}$, $\mathrm{rank}_{\mathbb{R}}(A; B) \leq 3$ if and only if $(A; B) = (O; O)$.

Suppose that A and B are linearly independent. In particular, A and B are nonzero and then nonsingular. $(A; B)$ is $\mathrm{GL}(4, \mathbb{R}) \times \{E_4\}$-equivalent to $(Q_1; A^{-1}B)$. $A^{-1}B$ is described as $x_1 Q_1 + x_2 Q_2 + x_3 Q_3 + x_4 Q_4$. Let $N = A^{-1}B - x_1 Q_1$. We see that $Q_j^{\top} = -Q_j$ and $Q_j^2 = -Q_1$ for $j = 2, 3, 4$, and $Q_i Q_j = -Q_j Q_i$ for $i, j = 2, 3, 4$ with $i \neq j$. Then, $N^{\top} = -N$ and $N^2 = -(x_2^2 + x_3^2 + x_4^2)Q_1$. For an eigenvalue λ of N, λ^2 is an eigenvalue of N^2. Since $N^2 = -N^{\top}N$, λ is equal to $\pm a\sqrt{-1}$, where $a = \sqrt{x_2^2 + x_3^2 + x_4^2}$. Therefore, the characteristic polynomial of N is equal to $(\lambda^2 + a^2)^2$. Since N is normal, N is diagonalizable over $\mathbb{C}$. Thus, there exists $P \in \mathrm{GL}(4, \mathbb{R})$ such that $P^{-1}N(a^{-1}P) = \frac{1}{a}C_2(0, a) = Q_2$, and then $(Q_1; N)$ is $\mathrm{GL}(4, \mathbb{R})^{\times 2}$-equivalent to $(Q_1; Q_2)$. Therefore, $(A; B)$ is $\mathrm{GL}(4, \mathbb{R})^{\times 2} \times \mathrm{GL}(2, \mathbb{R})$-equivalent to $(Q_1; Q_2)$ and $\mathrm{rank}_{\mathbb{R}}(A; B) = \mathrm{rank}_{\mathbb{R}}(Q_1; Q_2) = 6$ by Theorem 5.3.

Finally, suppose that $\mathrm{rank}_{\mathbb{R}}(A; B) = 4, 5$. A and B are linearly dependent and A or B is nonsingular. If A is nonsingular, then $B = yA$ for some $y \in \mathbb{R}$, and then $\mathrm{rank}_{\mathbb{R}}(A; B) = \mathrm{rank}_{\mathbb{R}}(A; O) = \mathrm{rank}(A) = 4$.

5.2 Upper Bound of the Maximal Rank of $m \times n \times 3$ Tensors

Kruskal (1977) studied the rank of a p-tensor and mainly obtained its lower bound. Atkinson and Stephens (1979) and Atkinson and Lloyd (1980) developed a nonlinear theory based on several of their own lemmas. Basically, they estimated the upper bound by adding two rectangular diagonal matrices, which allows the two matrices to be rectangular diagonalizable simultaneously. They did not solve the problem fully and restricted the type of tensors for obtaining clear-cut results.

Clearly,

$$\max.\mathrm{rank}_{\mathbb{K}}(m, n, mn) = mn.$$

Lemma 5.1 (Atkinson and Stephens 1979, Lemma 5) *Let $\mathbb{K}$ be a subfield of $\mathbb{F}$. If $k \leq n$, then*

$$\text{max.rank}_{\mathbb{K}}(m, n, mn - k) = m(n - k) + \text{max.rank}_{\mathbb{K}}(m, k, mk - k).$$

Proof Let $(A_1; \ldots; A_{mk-k}) \in T_{\mathbb{K}}(m, k, mk - k)$ be a tensor of rank

$$\text{max.rank}_{\mathbb{F}}(m, k, mk - k)$$

and $B_j = (A_j, O)$ be an $m \times n$ matrix for $1 \leq j \leq mk - k$. Consider the tensor $X = (B_1; \ldots; B_{mk-k}; E_{1,k+1}; \ldots; E_{1n}; \ldots; E_{m,k+1}; \ldots; E_{mn})$ with format $(m, n, mn - k)$, where E_{ij} denotes an $m \times n$ matrix with a 1 in the (i, j) position and zeros elsewhere. We have

$$
\begin{aligned}
\text{max.rank}_{\mathbb{K}}(m, n, mn - k) &\geq \text{rank}_{\mathbb{K}}(X) \\
&\geq \text{rank}_{\mathbb{K}}(B_1; \ldots; B_{mk-k}) + \text{rank}_{\mathbb{K}}(E_{1,k+1}; \ldots; E_{mn}) \\
&= \text{max.rank}_{\mathbb{K}}(m, k, mk - k) + m(n - k)
\end{aligned}
$$

(see Theorem 1.2).

We prove that

$$\text{max.rank}_{\mathbb{K}}(m, n, mn - k) \leq m(n - k) + \text{max.rank}_{\mathbb{K}}(m, k, mk - k).$$

Let $(A_1; \ldots; A_{mn-k}) \in T_{\mathbb{K}}(m, n, mn - k)$ be any tensor. To discuss an upper bound of the maximal rank, we may assume that $A_1, \ldots, A_{mn-k}$ are linearly independent without loss of generality. Let X be the vector space spanned by $m \times n$ matrices $A_1, \ldots, A_{mn-k}$. It suffices to show that X is a vector subspace of a vector space spanned by $m(n - k) + \text{max.rank}_{\mathbb{K}}(m, k, mk - k)$ rank-1 matrices. For $1 \leq i \leq m$, let Y_i be the vector space consisting of all matrices such that the i'th row is zero if $i' \neq i$. Since

$$\dim(X \cap Y_i) = \dim(X) + \dim(Y_i) - \dim(X \cap Y_i) \geq (mn - k) + n - mn = n - k,$$

we can take linearly independent matrices $B_1^{(i)}, \ldots, B_{n-k}^{(i)}$ of $X \cap Y_i$. Let $Z_1, \ldots, Z_{(m-1)k}$ be linearly independent matrices such that

$$X = \langle Z_1, \ldots, Z_{(m-1)k}, B_j^{(i)} \mid 1 \leq i \leq m, 1 \leq j \leq n - k \rangle,$$

and $V_i = \langle b_1^{(i)}, \ldots, b_{n-k}^{(i)} \rangle$, where $b_j^{(i)}$ is the ith row of $B_j^{(i)}$. There exists a k-dimensional vector subspace U of $\mathbb{K}^{1 \times n}$ such that $U + V_i = \mathbb{K}^{1 \times n}$ for arbitrary i. For $1 \leq u \leq (m - 1)k$, let $Z_u' = Z_u + \sum \alpha_{iju} B_j^{(i)}$ of which all rows are in U. Let

G be a nonsingular $n \times n$ matrix whose last $(n - k)$ columns lie in the orthogonal complement of U. Then,

$$\text{rank}_{\mathbb{K}}(Z_1'; \ldots; Z_{(m-1)k}') = \text{rank}_{\mathbb{K}}(Z_1'G; \ldots; Z_{(m-1)k}'G)$$
$$\leq \text{max.rank}_{\mathbb{K}}(m, k, (m-1)k),$$

since the last $(n - k)$ columns of the $m \times n$ matrix $Z_j'G$ are all zero. Therefore, there exist rank-1 matrices $C_1, \ldots, C_r$ such that $\langle Z_1', \ldots, Z_{(m-1)k}' \rangle \subset \langle C_1, \ldots, C_r \rangle$, and then X is spanned by $r + m(n - k)$ rank-1 matrices $C_1, \ldots, C_r, B_1^{(1)}, \ldots, B_{n-k}^{(m)}$, where $r = \text{max.rank}_{\mathbb{K}}(m, k, (m-1)k)$.

If $k = 1$, then

$$\text{max.rank}_{\mathbb{F}}(m, n, mn - 1) = m(n - 1) + (m - 1) = mn - 1.$$

Theorem 5.7 (Atkinson and Stephens 1979, Theorem 2) *Let $\mathbb{K}$ be a subfield of $\mathbb{F}$. If $k \leq m \leq n$, then*

$$\text{max.rank}_{\mathbb{K}}(m, n, mn - k) = mn - k^2 + \text{max.rank}_{\mathbb{K}}(k, k, k^2 - k).$$

Proof By Lemma 5.1, we see that

$$\text{max.rank}_{\mathbb{K}}(m, n, mn - k) = m(n - k) + \text{max.rank}_{\mathbb{K}}(m, k, mk - k)$$
$$= m(n - k) + \text{max.rank}_{\mathbb{K}}(k, m, mk - k)$$
$$= m(n - k) + k(m - k) + \text{max.rank}_{\mathbb{K}}(k, k, k^2 - k)$$
$$= mn - k^2 + \text{max.rank}_{\mathbb{K}}(k, k, k^2 - k).$$

If $k = 2$, then

$$\text{max.rank}_{\mathbb{F}}(m, n, mn - 2) = mn - 2^2 + \text{max.rank}_{\mathbb{F}}(2, 2, 2) = mn - 1.$$

Moreover, Atkinson and Lloyd (1983) studied the vector space of $m \times n$ matrices with dimension $mn - 2$ by applying the Kronecker–Weierstrass theory. They showed the following:

Theorem 5.8 (Atkinson and Lloyd 1983) *Let $A = (A_1; \ldots; A_{mn-2})$ be a tensor of $T_{\mathbb{F}}(m, n, mn - 2)$. Then, $\text{rank}_{\mathbb{F}}(A) = mn - 1$ if and only if A is $\text{GL}(m, \mathbb{F}) \times \text{GL}(n, \mathbb{F}) \times \text{GL}(mn - 2, \mathbb{F})$-equivalent to*

$$(E_{11} + E_{22}; B_2; \ldots; B_{mn-2}),$$

where $\{B_2, \ldots, B_{mn-2}\} = \{E_{ij} \mid (i, j) \neq (1, 1), (1, 2), (2, 2)\}$ and E_{ij} denotes an $m \times n$ matrix with a 1 in the (i, j) position and zeros elsewhere. If $A_1, \ldots, A_{mn-2}$ are linearly dependent, then $\text{rank}_{\mathbb{F}}(A) \leq mn - 2$, and in particular,

$$\text{max.rank}_{\mathbb{F}}(m, n, mn - 3) \leq mn - 2.$$

By Theorem 5.8, we have $\text{max.rank}_{\mathbb{F}}(3, 3, 6) \leq 7$. Indeed, $\text{max.rank}_{\mathbb{F}}(3, 3, 6) = 7$. It suffices to show that $\text{rank}_{\mathbb{K}}((E_3, O); (J, O); (O, E_3)) \geq 7$, where $J = \begin{pmatrix} 0 & 1 & 0 \\ 0 & 0 & 0 \\ 0 & 0 & 0 \end{pmatrix}$.
Suppose that $\text{max.rank}_{\mathbb{K}}(3, 3, 6) \leq 6$. There exist a 3×6 matrix P, a 6×6 matrix Q, and diagonal 6×6 matrices D_1, D_2, D_3 such that $(E_3, O) = PD_1Q$, $(J, O) = PD_2Q$ and $(O, E_3) = PD_3Q$. Then, $E_6 = \begin{pmatrix} PD_1 \\ PD_3 \end{pmatrix} Q$,

$$PD_2 = (J, O) \begin{pmatrix} PD_1 \\ PD_3 \end{pmatrix} = JPD_1 = \begin{pmatrix} P^{=2}D_1 \\ O \\ O \end{pmatrix},$$

where $P^{=2}$ denotes the second row vector of P. Thus, $P^{=2}D_2 = O$ from the second row, and then $PD_2^2 = O$. Since $D_2 \neq O$, if the (j, j) position of D_2 is not zero for some j, the jth column of P is a zero vector. Then, the jth column of $\begin{pmatrix} PD_1 \\ PD_3 \end{pmatrix}$ is also zero, which is a contradiction. We conclude that $\text{rank}_{\mathbb{K}}((E_3, O); (J, O); (O, E_3)) \geq 7$. Therefore,

$$\text{max.rank}_{\mathbb{F}}(m, n, mn - 3) = mn - 2$$

for $m, n \geq 3$ by Theorem 5.7.

Theorem 5.9 (Atkinson and Lloyd 1980, Theorem 4') *Let $A_1 = (E_m, O, O, \ldots, O)$, $A_2 = (O, E_m, O, \ldots, O), \ldots, A_{k-1} = (O, O, \ldots, O, E_m)$, and A_k be $m \times (k-1)m$ matrices. Then,*

$$\text{rank}_{\mathbb{F}}(A_1; A_2; \ldots; A_k) \leq mk - \left\lceil \frac{m}{2} \right\rceil.$$

Note that almost all tensors with format $(m, (k-1)m, k)$ are $\text{GL}(m, \mathbb{F}) \times \text{GL}((k-1)m, \mathbb{F}) \times \text{GL}(k, \mathbb{F})$-equivalent to $(A_1; \ldots; A_k)$ in the above theorem.

Conjecture 5.1 (Atkinson and Stephens 1979)

$$\text{max.rank}_{\mathbb{F}}(n, n, n^2 - n) = n^2 - \left\lceil \frac{n}{2} \right\rceil.$$

According to Atkinson and Stephens 1979, Lloyd showed that

$$\text{max.rank}_{\mathbb{F}}(m, n, mn - k) \geq mn - \left\lceil \frac{k}{2} \right\rceil$$

for $k \leq m \leq n$, which is unpublished. We consider this inequality. By Theorem 5.7, it suffices to show that $\text{max.rank}_{\mathbb{K}}(m, m, m^2 - m) \geq m^2 - \lceil m/2 \rceil$. By Theorem 5.9, *letting* $k = m$, *we see that* $\mathrm{rank}_{\mathbb{K}}(A_1; \ldots; A_m) = m(m-1) + \lfloor m/2 \rfloor = m^2 - \lceil m/2 \rceil$ *for some* A_m *(see also Theorem 5.19). Hence,*

$$
\begin{aligned}
\mathrm{max.rank}_{\mathbb{K}}(m, m, m^2 - m) &= \mathrm{max.rank}_{\mathbb{K}}(m, m^2 - m, m) \\
&\geq \mathrm{rank}_{\mathbb{K}}(A_1; \ldots; A_m) \\
&= m^2 - \lceil m/2 \rceil.
\end{aligned}
$$

Lemma 5.2 (cf. Atkinson and Stephens 1979, Lemma 4) *Let* $m \leq n$ *and A, B be* $m \times m$ *matrices. If A is nonsingular and* $A^{-1}B$ *is diagonalizable, then*

$$
\mathrm{rank}_{\mathbb{F}}((A, X); (B, Y)) \leq n
$$

for any $m \times (n - m)$ *matrices X and Y.*

We remark that the maximal rank of $T_{\mathbb{F}}(m, n, 2)$ is equal to $m + \lfloor n/2 \rfloor$. Then, this lemma is obtained as a conclusion if $n > 2m$.

Proof We may assume that $n \leq 2m$. Let P be a nonsingular $m \times m$ matrix such that $P^{-1}A^{-1}BP$ is a diagonal matrix. Let X, Y be $m \times (n - m)$ matrices and put $D = P^{-1}A^{-1}BP$, $X' = P^{-1}A^{-1}X$, and $Y' = P^{-1}A^{-1}Y$. Then, $((A, X); (B, Y))$ is $GL(m, \mathbb{F}) \times GL(n, \mathbb{F})$-equivalent to $((E_m, O); (D, Y' - DX'))$. Therefore,

$$
\mathrm{rank}_{\mathbb{F}}((A, X); (B, Y)) \leq \mathrm{rank}_{\mathbb{F}}(E_m; D) + \mathrm{rank}(Y' - DX') \leq m + (n - m) = n.
$$

Remark 5.1 Let A and B be $m \times m$ matrices over $\mathbb{F}$. For sufficiently large $s \in \mathbb{F}$,

$$
(A + s\mathrm{Diag}(1, 2, \ldots, m))^{-1}(B + sE_m) = \left(\frac{1}{s}A + \mathrm{Diag}(1, 2, \ldots, m) \right)^{-1} \left(\frac{1}{s}B + E_m \right)
$$

has distinct eigenvalues, which are all real if $\mathbb{F} = \mathbb{R}$.

By this remark, we obtain the following proposition:

Proposition 5.3 *Let* $3 \leq m \leq n$ *and* $T = ((E_m, O); A; B) \in T_{\mathbb{F}}(m, n, 3)$. *Then,*

$$
\mathrm{rank}_{\mathbb{F}}(T) \leq m + n.
$$

Theorem 5.10 (Atkinson and Stephens 1979, Theorem 4; Sumi et al. 2010, Theorems 5, 6)

(1) $\mathrm{max.rank}_{\mathbb{C}}(n, n, 3) \leq 2n - 1$ *and* $\mathrm{max.rank}_{\mathbb{R}}(n, n, 3) \leq 2n$.
(2) *If* $m < n$, *then* $\mathrm{max.rank}_{\mathbb{F}}(m, n, 3) \leq m + n - 1$.
(3) *If* n *is not congruent to* 0 *modulo* 4, *then* $\mathrm{max.rank}_{\mathbb{R}}(n, n, 3) \leq 2n - 1$.

Proof The proof of (1) and (2) is seen in Sumi et al. 2010, Theorem 5 and Sumi et al. 2010, Theorem 6, respectively. For the proof of (3), in the proof of Sumi et al.

2010, Theorem 5 over $\mathbb{R}$, the assumption that n is odd only uses the fact that for a tensor $(A; B; C)$ with format $(n, n, 3)$ the vector space generated by the slices A, B, and C contains a singular matrix. Now, we know that this is equivalent to the fact that $n \not\equiv 0$ modulo 4. Therefore, we proceed with the proof of Sumi et al. 2010, Theorem 5 over $\mathbb{R}$.

Corollary 5.1 max.rank$_{\mathbb{R}}(3, 3, 3) \le 5$.

In Chap. 3, we see that max.rank$_{\mathbb{R}}(3, 3, 3) = 5$. To conclude this section, we show that max.rank$_{\mathbb{R}}(4, 4, 3) \ge 7$.

$$
\text{Let } Q_1 = E_4,\ Q_2 = \begin{pmatrix} & & -1 & \\ 1 & & & \\ & & & -1 \\ & 1 & & \end{pmatrix},\ Q_3 = \begin{pmatrix} & & -1 & \\ & & & 1 \\ 1 & & & \\ & -1 & & \end{pmatrix},\ Q_4 = \begin{pmatrix} & & & 1 \\ & & 1 & \\ & -1 & & \\ -1 & & & \end{pmatrix},
$$

and let $W = \langle Q_1, Q_2, Q_3, Q_4 \rangle$ be a real vector space. Let M_1, M_2, M_3 be linearly independent matrices of W. Suppose that rank$_{\mathbb{R}}(M_1; M_2; M_3) \le 6$ as a contradiction. There exist rank-1 matrices $C_1, \ldots, C_6$ and $c_{ij} \in \mathbb{R}$, $1 \le i \le 3$ and $1 \le j \le 6$, such that $M_i = \sum_{j=1}^{6} c_{ij} C_j$. We may assume that $c_{36} \ne 0$ without loss of generality. Put $N_i = M_i - \frac{c_{i6}}{c_{36}} M_3 \in W$ for $i = 1, 2$. Clearly, N_1 and N_2 are linearly independent, $\langle N_1, N_2 \rangle \subset \langle C_1, \ldots, C_5 \rangle$, and thus rank$_{\mathbb{R}}(N_1; N_2) \le 5$. This is a contradiction by Proposition 5.2. Therefore, rank$_{\mathbb{R}}(M_1; M_2; M_3) \ge 7$.

5.3   Maximal Rank of Higher Tensors

In this section, we consider more fundamental properties for the maximal rank of 3-tensors. Since an n-tensor is a collection of $(n - 1)$-tensors, we use them simultaneously. In particular, for a 3-tensor, we can apply matrix theory simultaneously. Let $\mathbb{K}$ be a field and let $T_{\mathbb{K}}(f_1, \ldots, f_n)$ be the set of all tensors with format $(f_1, \ldots, f_n)$ whose elements are in $\mathbb{K}$.

Theorem 5.11 *Let $1 \le m \le n \le p$. Let T be a 3-tensor with format $f = (m, n, p)$, $C_1, \ldots, C_t$ be rank-1 tensors with format f, and $T + C_1 + \cdots + C_t = (A_1; \ldots; A_p)$. If $m \times n$ tensors $A_1, A_2, \ldots, A_p$ are simultaneously rectangular diagonalizable, then* rank$_{\mathbb{K}}(T) \le m + t$.

Proof If $A_1, A_2, \ldots, A_p$ are simultaneously rectangular diagonalizable, then

$$
\text{rank}_{\mathbb{K}}(A_1; A_2; \ldots; A_p) \le m.
$$

Thus,

$$
\text{rank}_{\mathbb{K}}(T) \le \text{rank}_{\mathbb{K}}(A_1; A_2; \ldots; A_p) + \sum_{k=1}^{t} \text{rank}_{\mathbb{K}}(C_k) \le m + t.
$$

For $0 \leq s \leq \text{max.rank}_{\mathbb{K}}(f)$, there exists a tensor $A \in \mathbb{K}^f$ with $\text{rank}_{\mathbb{K}}(A) = s$.

Let $A = (a_{i,j,k}) = (A_1; A_2; \ldots; A_{f_3})$ be a tensor with format (f_1, f_2, f_3) such that $a_{f_1, f_2, k} = 1$ for any k with $1 \leq k \leq f_3$. For $k = 1, 2, \ldots, f_3$, let u_k (resp. v_k) be the f_1th row (resp. f_2th column) of the $f_1 \times f_2$ matrix A_k. Then, all elements of the f_1th row and f_2th column of $A_k - v_k u_k$ are zero. Then,

$$\text{rank}_{\mathbb{K}}(A) \leq \text{rank}_{\mathbb{K}}(A_1 - v_1 u_1; \ldots; A_{f_3} - v_{f_3} u_{f_3}) + \text{rank}_{\mathbb{K}}(v_1 u_1; \ldots; v_{f_3} u_{f_3})$$
$$\leq \text{max.rank}_{\mathbb{K}}(f_1 - 1, f_2 - 1, f_3) + f_3.$$

Proposition 5.4 (cf. Howell 1978, Theorem 8) *Suppose that $\mathbb{K}$ is an infinite field. Let $k \geq 3$.*

$$\text{max.rank}_{\mathbb{K}}(f_1, f_2, f_3, \ldots, f_k) \leq \prod_{i=3}^{k} f_i + \text{max.rank}_{\mathbb{K}}(f_1 - 1, f_2 - 1, f_3, \ldots, f_k).$$

Proof Let $X = (x_{i_1, i_2, \ldots, i_k}) \in T_{\mathbb{K}}(f_1, f_2, \ldots, f_k) \setminus \{O\}$. There exists $Y = (y_{i_1, i_2, \ldots, i_k}) \in T_{\mathbb{K}}(f_1, f_2, \ldots, f_k)$ such that Y is equivalent to X and

$$y_{f_1, f_2, i_3, \ldots, i_k} \neq 0$$

for any $i_3, \ldots, i_k$. For each $(j_3, \ldots, j_k)$ with $1 \leq j_t \leq f_t$, $3 \leq t \leq k$, we consider the rank-1 tensor $A_{j_3, \ldots, j_k}$ defined as

$$\frac{1}{y_{f_1, f_2, j_3, \ldots, j_k}} \begin{pmatrix} y_{1, f_2, j_3, \ldots, j_k} \\ y_{2, f_2, j_3, \ldots, j_k} \\ \cdots \\ y_{f_1, f_2, j_3, \ldots, j_k} \end{pmatrix} \otimes \begin{pmatrix} y_{f_1, 1, j_3, \ldots, j_k} \\ y_{f_1, 2, j_3, \ldots, j_k} \\ \cdots \\ y_{f_1, f_2, j_3, \ldots, j_k} \end{pmatrix} \otimes e_{j_3}^{(3)} \otimes e_{j_4}^{(4)} \otimes \cdots \otimes e_{j_k}^{(k)},$$

where $e_j^{(t)}$ is the jth column vector of the $f_t \times f_t$ identity matrix. Put $X' = X - \sum_{j_3, \ldots, j_k} A_{j_3, \ldots, j_k}$. The $(i_1, f_2, i_3, \ldots, i_k)$th and $(f_1, i_2, i_3, \ldots, i_k)$th elements of X' are all zero for any $i_1, i_2, i_3, \ldots, i_k$. Thus,

$$\text{rank}_{\mathbb{K}}(X') \leq \text{max.rank}_{\mathbb{K}}(f_1 - 1, f_2 - 1, f_3, \ldots, f_k),$$

and then,

$$\text{rank}_{\mathbb{K}}(X) = \text{rank}_{\mathbb{K}}(Y) \leq \prod_{i=3}^{k} f_i + \text{max.rank}_{\mathbb{K}}(f_1 - 1, f_2 - 1, f_3, \ldots, f_k).$$

Bailey and Rowley (1993) obtained a similar result for 3-tensors.

Theorem 5.12 (Howell 1978, Theorem 7) *Suppose that $\mathbb{K}$ is an infinite field.*

$$\text{max.rank}_{\mathbb{K}}(n, n, n) \leq \lceil 3n^2/4 \rceil.$$

Proof If we apply Proposition 5.4 $\lceil n/2 \rceil$ times for $k = 3$, then

$$\text{max.rank}_{\mathbb{K}}(n, n, n) \leq n\lceil n/2 \rceil + \text{max.rank}_{\mathbb{K}}(\lfloor n/2 \rfloor, \lfloor n/2 \rfloor, n)$$
$$\leq n\lceil n/2 \rceil + \lfloor n/2 \rfloor^2$$
$$= \lceil 3n^2/4 \rceil.$$

Theorem 5.13 (cf. Howell 1978, Theorem 9) *Suppose that* $\mathbb{K}$ *is an infinite field.*

$$\text{max.rank}_{\mathbb{K}}(\overbrace{n, n, \ldots, n}^{d}) \leq \frac{d}{2(d-1)} n^{d-1} + o(n^{d-1}).$$

Proof If we apply Proposition 5.4 d times, then

$$\text{max.rank}_{\mathbb{K}}(\overbrace{n, n, \ldots, n}^{d}) \leq dn^{d-2} + \text{max.rank}_{\mathbb{K}}(\overbrace{n-2, n-2, \ldots, n-2}^{d})$$
$$= dn^{d-2} + d(n-2)^{d-2} + d(n-4)^{d-2} + \cdots$$
$$= \frac{d}{2(d-1)} n^{d-1} + o(n^{d-2}).$$

Theorem 5.14 (cf. Atkinson and Lloyd 1980, Theorem 1; Sumi et al. 2010, Theorem 3) *Let* $3 \leq m \leq n \leq p$. *Let* $p = 2q + \varepsilon$ *for an integer* q *and* $\varepsilon = 0, 1$.

$$\text{max.rank}_{\mathbb{F}}(m, n, p) \leq (\varepsilon + 1)m + \left\lfloor \frac{n(p-1-\varepsilon)}{2} \right\rfloor.$$

Proof Let $A = (A_1; A_2; \ldots; A_p)$. First, we assume that $\varepsilon = 0$. Let $n' = \lfloor n/2 \rfloor$. Since the maximal rank of $T_{\mathbb{F}}(m, n, 2)$ is equal to $m + n'$, there exists a tensor $(C_1; C_2) \in T_{\mathbb{F}}(m, n, 2)$, $P \in \text{GL}(m, \mathbb{F})$, and $Q \in \text{GL}(n, \mathbb{F})$ such that $\text{rank}_{\mathbb{F}}(C_1; C_2) \leq n'$, $PA_{p-1}Q = C_1 + (D_{p-1}, O)$, and $PA_pQ = C_2 + (D_p, O)$, where D_{p-1} and D_p are $m \times m$ diagonal matrices. Let $B = PAQ = (B_1; B_2; \ldots; B_p)$.

$$\text{rank}_{\mathbb{F}}(A) \leq \text{rank}_{\mathbb{F}}(B) + \text{rank}_{\mathbb{F}}(O; \ldots; O; C_1; C_2)$$

$$\leq \sum_{j=1}^{q-1} \text{rank}_{\mathbb{F}}(B_{2j-1} - (D_{2j-1}, O); B_{2j} - (D_{2j}, O))$$

$$+ \text{rank}_{\mathbb{F}}(D_1; D_2; \ldots; D_p) + \text{rank}_{\mathbb{F}}(C_1; C_2)$$

for any diagonal $m \times m$ matrices $D_1, D_2, \ldots, D_{p-2}$. By Lemma 5.2 and Remark 5.1, we have

$$\text{rank}_{\mathbb{F}}(B_{2j-1} - (D_{2j-1}, O); B_{2j} - (D_{2j}, O)) \leq n$$

for some D_{2j-1}, D_{2j} and any j with $1 \leq j \leq q - 1$. Then, we see that $\mathrm{rank}_{\mathbb{F}}(A) \leq n(q-1) + m + n' = m + \lfloor \frac{n(p-1)}{2} \rfloor$.

If p is odd, then by the above estimation,

$$\mathrm{rank}_{\mathbb{F}}(A) = \mathrm{rank}_{\mathbb{F}}(A_1; A_2; \ldots; A_{p-1}) + \mathrm{rank}(A_p)$$
$$\leq 2m + \left\lfloor \frac{n(p-2)}{2} \right\rfloor.$$

For a subset of $\mathscr{U}$ of $T_{\mathbb{R}}(f)$, cl $\mathscr{U}$ and int $\mathscr{U}$ denote the Euclidean closure and interior of $\mathscr{U}$, respectively.

The border rank (denoted by $\mathrm{brank}_{\mathbb{K}}(T)$) of a tensor T over $\mathbb{K}$ is the minimal integer r such that there exist tensors T_n, $n \geq 1$ with rank r that converge to T as n goes to infinity. By definition, we see that $\mathrm{brank}_{\mathbb{K}}(T) \leq \mathrm{rank}_{\mathbb{K}}(T)$.

Theorem 5.15 (Strassen 1983, Theorem 4.1) *Let $\mathbb{K}$ be an algebraically closed field or $\mathbb{R}$, and $(A; B; C) \in T_{\mathbb{K}}(n, n, 3)$. If A is nonsingular, then*

$$\mathrm{brank}_{\mathbb{K}}(A; B; C) \geq n + \frac{1}{2}\mathrm{rank}(BA^{-1}C - CA^{-1}B).$$

Proof First, we show that $\mathrm{rank}_{\mathbb{K}}(A; B; C) \geq n + \frac{1}{2}\mathrm{rank}(BA^{-1}C - CA^{-1}B)$. The tensor $(A; B; C)$ is $\mathrm{GL}(n, \mathbb{K}) \times \{E_n\} \times \{E_3\}$-equivalent to $(E_n; A^{-1}B; A^{-1}C)$. Let $B' = A^{-1}B$, $C' = A^{-1}C$, and $r = \mathrm{rank}_{\mathbb{K}}(E_n; B'; C') = \mathrm{rank}_{\mathbb{K}}(A; B; C)$. There exist an $n \times r$ matrix P, $r \times r$ diagonal matrices D_1, D_2, D_3, and an $r \times n$ matrix Q such that $E_n = PD_1Q$, $B' = PD_2Q$, and $C' = PD_3Q$. Note that the vector space $\langle D_1, D_2, D_3 \rangle$ of diagonal matrices has a nonsingular matrix. We show that $\mathrm{rank}(B'C' - C'B') \leq 2(r - n)$ by separating into two cases.

Suppose that D_1 is nonsingular. Put $Q' = D_1Q, D_2' = D_2D_1^{-1}$, and $D_3' = D_3D_1^{-1}$. Then, $E_n = PQ', B' = PD_2'Q'$, and $C' = PD_3'Q'$. Note that the rows of P are linearly independent and so are the columns of Q', since $PQ' = E_n$. Let $\hat{Q}$ be a nonsingular $r \times r$ matrix obtained from Q' by attaching $(r - n)$ columns to the right-hand side orthogonal to the rows of P and put $\hat{P} = \hat{Q}^{-1}$. Since $P\hat{Q} = (A', O)$, $\hat{P}$ is obtained from P by attaching $(r - n)$ rows to the bottom. For $\hat{B} = \hat{P}D_2'\hat{Q}$ and $\hat{C} = \hat{P}D_3'\hat{Q}$, we can write

$$\hat{B} = \begin{pmatrix} B' & B_{12} \\ B_{21} & B_{22} \end{pmatrix}, \quad \hat{C} = \begin{pmatrix} C' & C_{12} \\ C_{21} & C_{22} \end{pmatrix}.$$

Since $\hat{B}$ commutes with $\hat{C}$, we see that

$$B'C' - C'B' = C_{12}B_{21} - B_{12}C_{21},$$

and then $\mathrm{rank}(B'C' - C'B') \leq 2(r - n)$, since B_{12} and C_{12} have only $(r - n)$ columns.

Suppose that D_1 is singular. If necessary we exchange P, Q, D_1, D_2, and D_3, we may assume that $D_1 = \mathrm{Diag}(a_1, \ldots, a_s, 0, \ldots, 0)$ with $a_1, \ldots, a_s \neq 0$ without loss of generality. Let $D_k = \mathrm{Diag}(D_{k1}, D_{k2})$ and $D_{k1}' = D_{k1}D_{11}^{-1}$ for $k = 1, 2, 3$,

$P = (P_1, P_2)$, $Q = \begin{pmatrix} Q_1 \\ Q_2 \end{pmatrix}$, $Q_1' = D_{11}Q_1$, and $Q' = (Q_1', Q_2)$, where D_{kl} is an $s \times s$ matrix, P_1 is an $n \times s$ matrix, and Q_1 is an $s \times n$ matrix. Then, $E_n = P_1 Q_1'$, $B' = P_1 D_{21}' Q_1' + P_2 D_{22} Q_2$, and $C' = P_1 D_{31}' Q_1' + P_2 D_{32} Q_2$. Put $B'' = P_1 D_{21}' Q_1'$ and $C'' = P_1 D_{31}' Q_1'$. Note that

$$B'C' - C'B' = (B''C'' - C''B'') + R_1 Q_2 + P_2 R_2$$

for some matrices R_1 and R_2 of appropriate size. We apply the above argument for (P_1, Q_1', B'', C'') instead of (P, Q', B', C'). Then, $\text{rank}(B''C'' - C''B'') \leq 2(s - n)$. Since $\text{rank}P_2 \leq r - s$ and $\text{rank}Q_2 \leq r - s$, we see that $\text{rank}(B'C' - C'B') \leq 2(s - n) + (r - s) + (r - s) = 2(r - n)$.

Therefore,

$$r \geq n + \frac{1}{2}\text{rank}(BA^{-1}C - CA^{-1}B),$$

since $B'C' - C'B' = A^{-1}(BA^{-1}C - CA^{-1}B)$.

Now, we show the assertion of the theorem. Let $r \geq n$ be an integer and let

$N_1(r) = \{X \in T_{\mathbb{K}}(n, n, 3) \mid \text{brank}_{\mathbb{K}}(X) \leq r\}$,
$N_2(r) = \{X \in T_{\mathbb{K}}(n, n, 3) \mid \text{rank}_{\mathbb{K}}(X) \leq r\}$,
$N_3(r) = \{(P; Q; R) \in T_{\mathbb{K}}(n, n, 3) \mid$
$\qquad\qquad \text{rank}(P) = n, \ n + \frac{1}{2}\text{rank}(QP^{-1}R - RP^{-1}Q) \leq r\}$,
$N_4 = \{(P; Q; R) \in T_{\mathbb{K}}(n, n, 3) \mid \text{rank}(P) < n\}$,
$N_5(r) = \{(P; Q; R) \in T_{\mathbb{K}}(n, n, 3) \mid n + \frac{1}{2}\text{rank}(Q\,\text{adj}(P)R - R\,\text{adj}(P)Q) \leq r\}$,

where $\text{adj}(P)$ denotes the adjoint of P. Then, $N_2(r) \subset N_3(r) \cup N_4 \subset N_5(r) \cup N_4$. Since $N_5(r) \cup N_4$ is closed, $N_1(r) = \text{cl}\,N_2(r) \subset N_5(r) \cup N_4$. For a sufficiently small open neighborhood U of $(A; B; C)$, P is nonsingular for any tensor $(P; Q; R) \in U$. If $r \geq \text{brank}_{\mathbb{K}}(A; B; C)$, then $N_1(r) \cap U \subset N_5(r) \cap U = N_3(r) \cap U$. Taking $k = \text{brank}_{\mathbb{K}}(A; B; C)$, since $(A; B; C) \in N_1(k) \cap U \subset N_3(k)$,

$$n + \frac{1}{2}\text{rank}(BA^{-1}C - CA^{-1}B) \leq \text{brank}_{\mathbb{K}}(A; B; C).$$

Theorem 5.16 *Let $A = (A_1; \ldots; A_{f_n}) \in T_{\mathbb{K}}(f_1, \ldots, f_n)$ and $1 \leq u < f_n$.*

$$\text{rank}_{\mathbb{K}}(A) \geq \min_{g = (*, E_{f_n - u}) \in T_{\mathbb{K}}(f_n - u, f_n)} \text{rank}_{\mathbb{K}}(A \times_n g) + \dim\langle A_1, A_2, \ldots, A_u \rangle.$$

If $P = (Q, E_s)$, then $A \times_n P$ is obtained from the first s slices of $A \times_n \tilde{P}$ for $\tilde{P} = \begin{pmatrix} Q & E_s \\ E_{f_n - s} & O \end{pmatrix} \in \text{GL}(f_n, \mathbb{K})$.

Proof (of Theorem 5.16) Let $r = \mathrm{rank}_{\mathbb{K}}(A)$. First, we show the inequality in the case where $A_1, \ldots, A_u$ are linearly independent. Suppose that $A_1, \ldots, A_u$ are linearly independent. Then, $r \geq u$, and there exist rank-1 tensors $C_1, \ldots, C_r$ with format $(f_1, \ldots, f_{n-1})$ such that

$$A_1, \ldots, A_{f_n} \in \langle C_1, \ldots, C_r \rangle \text{ and } C_1, \ldots, C_u \in \langle A_1, \ldots, A_u, C_{u+1}, \ldots, C_r \rangle.$$

We write

$$A_k = \sum_{i=1}^{r} \alpha_{ki} C_i \quad (1 \leq k \leq f_n) \text{ and}$$

$$C_j = \sum_{h=1}^{u} \beta_{jh} A_h + \sum_{h=1}^{r-u} \gamma_{jh} C_{h+u} \quad (1 \leq j \leq u)$$

for some $(\alpha_{ki})_{k,i} \in T_{\mathbb{K}}(f_n, r)$, $(\beta_{jh})_{j,h} \in T_{\mathbb{K}}(u, u)$, $(\gamma_{jh})_{j,h} \in T_{\mathbb{K}}(u, r-u)$. We see that

$$A_k - \sum_{h=1}^{u} \left(\sum_{j=1}^{u} \alpha_{kj} \beta_{jh} \right) A_h = \sum_{h=u+1}^{r} \left(\alpha_{kh} + \sum_{j=1}^{u} \alpha_{kj} \gamma_{j,h-u} \right) C_h$$

for $u + 1 \leq k \leq f_n$. Put $P = (\alpha_{ki})_{u<k\leq f_n, 1\leq i \leq u} \in T_{\mathbb{K}}(f_n - u, u)$, $Q = (\beta_{jh})_{j,h} \in T_{\mathbb{K}}(u, u)$, and $g_0 = (-PQ, E_{f_n-u})$. Since

$$A'_k := A_k - \sum_{h=1}^{u} \left(\sum_{j=1}^{u} \alpha_{kj} \beta_{jh} \right) A_h \in \langle C_{u+1}, \ldots, C_r \rangle$$

for $1 \leq k \leq u$ and $C_{u+1}, \ldots, C_r$ have rank 1, we see that

$$(A'_{u+1}; \ldots; A'_{f_n}) = A \times_n g_0 \quad \text{and} \quad \mathrm{rank}_{\mathbb{K}}(A'_{u+1}; \ldots; A'_{f_n}) \leq r - u.$$

Thus, $r \geq \min_{g} \mathrm{rank}_{\mathbb{K}}(A \times_n g) + u$.

We consider the case where $A_1, \ldots, A_u$ are linearly dependent. Let $B_1, B_2, \ldots, B_v$ be a basis of $\langle A_1, A_2, \ldots, A_u \rangle$. Then, we have the equality $\mathrm{rank}_{\mathbb{K}}(A) = \mathrm{rank}_{\mathbb{K}}(B)$ for $B := (B_1; \ldots; B_v; A_{u+1}; \ldots; A_{f_n})$. By applying the above argument for B, we have

$$\mathrm{rank}_{\mathbb{K}}(B) \geq \min_{h=(*, E_{f_n-u})} \mathrm{rank}_{\mathbb{K}}(B \times_n h) + v.$$

For any matrix $h = (H, E_{f_n-u}) \in T_{\mathbb{K}}(v, f_n + v - u)$, there exists $g = (G, E_{f_n-u}) \in T_{\mathbb{K}}(u, f_n - u)$ such that $B \times_n h = A \times_n g$. Therefore,

$$\mathrm{rank}_{\mathbb{K}}(A) \geq \min_{g=(*, E_{f_n-u})} \mathrm{rank}_{\mathbb{K}}(A \times_n g) + v.$$

For a permutation σ on n letters, let

$$\varphi_\sigma : T_{\mathbb{K}}(f_1, \ldots, f_n) \to T_{\mathbb{K}}(f_{\sigma(1)}, \ldots, f_{\sigma(n)})$$

be a bijection sending $(a_{i_1,\ldots,i_n})_{i_1,\ldots,i_n}$ to $(a_{i_1,\ldots,i_n})_{i_{\sigma(1)},\ldots,i_{\sigma(n)}}$. By the definition of rank, we see that $\mathrm{rank}_{\mathbb{K}}(A) = \mathrm{rank}_{\mathbb{K}}(\varphi_\sigma(A))$.

Proposition 5.5 *Let $A \in T_{\mathbb{K}}(f_1, \ldots, f_n)$, $1 \le t \le n$, $1 \le s \le f_t$ and $Q \in T_{\mathbb{K}}(s, f_t)$. It holds that $\varphi_\sigma(A) \times_t Q = \varphi_\sigma(A \times_{\sigma^{-1}(t)} Q)$.*

Proof The proof is omitted as it is straightforward.

For $A = (A_1; \ldots; A_k) \in T_{\mathbb{K}}(m, n, k)$, the column rank $\mathrm{col\_rank}(A)$ is defined as the rank of the $mk \times n$ matrix $\mathrm{fl}_2(A)$ and the row rank $\mathrm{row\_rank}(A)$ is defined as the rank of the $m \times nk$ matrix $\mathrm{fl}_1(A)$.

Proposition 5.6 *For $A \in T_{\mathbb{K}}(f_1, f_2, f_3)$, let*

$$(B_1; \ldots; B_{f_2}) = \varphi_{(1,3,2)}(A) \text{ and } (C_1; \ldots; C_{f_1}) = \varphi_{(1,2,3)}(A).$$

Then, $\mathrm{col\_rank}(A) = \dim\langle B_1, \ldots, B_{f_2}\rangle$ and $\mathrm{row\_rank}(A) = \dim\langle C_1, \ldots, C_{f_1}\rangle$.

Proof Let $A = (a_{ijk})$.

Recall that $\dim\langle N_1, \ldots, N_u\rangle = \mathrm{rank}(vec(N_1), \ldots, vec(N_u))$ for $N_1, \ldots, N_u \in T_{\mathbb{K}}(s, t)$. Then,

$$\mathrm{row\_rank}(A) = \mathrm{rank}((a_{1jk}), \ldots, (a_{f_1 jk})) = \dim\langle C_1, \ldots, C_{f_1}\rangle,$$

since $C_s = (a_{sjk})_{j,k}$ for $1 \le s \le f_1$. Similarly, we see that

$$\mathrm{col\_rank}(A) = \dim\langle B_1, \ldots, B_{f_2}\rangle.$$

By applying Theorem 5.16 for a 3-tensor and changing the slice direction, we have the following result.

Theorem 5.17 (Brockett and Dobkin 1978, Theorem 9) *Let $1 \le s < m$, $1 \le t < n$, $1 \le u < p$, and $A = (A_1; A_2; \ldots; A_p) \in T_{\mathbb{K}}(m, n, p)$. Let $A_j = (B_j, C_j) = \begin{pmatrix} P_j \\ Q_j \end{pmatrix}$, $j = 1, \ldots, p$, where B_j is an $m \times t$ matrix and P_j is an $s \times n$ matrix. Then, $\mathrm{rank}_{\mathbb{K}}(A)$ is greater than or equal to the following numbers:*

(1)

$$\min_{(a_{ij})} \mathrm{rank}_{\mathbb{K}}\left(A_1 + \sum_{j=u+1}^{p} a_{1j}A_j; A_2 + \sum_{j=u+1}^{p} a_{2j}A_j; \ldots; A_u + \sum_{j=u+1}^{p} a_{uj}A_j\right)$$
$$+ \dim\langle A_{u+1}, A_{u+2}, \ldots, A_p\rangle$$

(2)

$$\min_{M \in T_{\mathbb{K}}(n-t,t)} \mathrm{rank}_{\mathbb{K}}(B_1 + C_1 M; B_2 + C_2 M; \ldots; B_p + C_p M)$$
$$+ \mathrm{col\_rank}(C_1; C_2; \ldots; C_p)$$

(3)

$$\min_{N \in T_{\mathbb{K}}(s,m-s)} \mathrm{rank}_{\mathbb{K}}(P_1 + N Q_1; P_2 + N Q_2; \ldots; P_p + N Q_p)$$
$$+ \mathrm{row\_rank}(Q_1; Q_2; \ldots; Q_p)$$

Proof (1) Let $A' = (A'_1; \ldots; A'_p) = (A_{u+1}; \ldots; A_p; A_1; \ldots, A_u)$. By applying Theorem 5.16 for the 3-tensor A', we have

$$\mathrm{rank}_{\mathbb{K}}(A) \geq \min_{g=(*,E_u)} \mathrm{rank}_{\mathbb{K}}(A' \times_3 g) + \dim\langle A'_1, A'_2, \ldots, A'_{p-u}\rangle$$
$$= \min_{g=(E_u,*)} \mathrm{rank}_{\mathbb{K}}(A \times_3 g) + \dim\langle A_{u+1}, A_{u+2}, \ldots, A_p\rangle$$

For $g = (E_u, (a_{ij}))$, we see that

$$A \times_3 g = \left(A_1 + \sum_{j=u+1}^{p} a_{1,j-u} A_j; \ldots; A_u + \sum_{j=u+1}^{p} a_{u,j-u} A_j \right).$$

Next, we consider (2) and (3). We see that

$$\mathrm{rank}_{\mathbb{K}}(\varphi_\sigma(A)) \geq \min_{g=(E_u,*)} \mathrm{rank}_{\mathbb{K}}(\varphi_\sigma(A) \times_3 g) + \dim\langle X_{u+1}, X_{u+2}, \ldots\rangle,$$

where $(X_1; X_2; \ldots) = \varphi_\sigma(A)$. Then,

$$\mathrm{rank}_{\mathbb{K}}(A) \geq \min_{g=(E_u,*)} \mathrm{rank}_{\mathbb{K}}(\varphi_\sigma(A \times_{\sigma^{-1}(3)} g)) + \dim\langle X_{u+1}, X_{u+2}, \ldots\rangle$$
$$= \min_{g=(E_u,*)} \mathrm{rank}_{\mathbb{K}}(A \times_{\sigma^{-1}(3)} g) + \dim\langle X_{u+1}, X_{u+2}, \ldots\rangle,$$

by Proposition 5.5. By Proposition 5.6, if $\sigma = (1, 2, 3)$ and $u = t$, then

$$\mathrm{rank}_{\mathbb{K}}(A) \geq \min_{g=(E_t,M)} \mathrm{rank}_{\mathbb{K}}(A \times_2 g) + \mathrm{row\_rank}(Q_1; \ldots; Q_p),$$

and if $\sigma = (1, 3, 2)$ and $u = s$, then

$$\mathrm{rank}_{\mathbb{K}}(A) \geq \min_{g=(E_s,N)} \mathrm{rank}_{\mathbb{K}}(A \times_1 g) + \mathrm{col\_rank}(C_1; \ldots; C_p).$$

Theorem 5.18 (Brockett and Dobkin 1978, Theorem 10) *For a tensor $A = (A_1; A_2; \ldots; A_p) \in T_{\mathbb{K}}(m, n, p)$, we have the following inequality.*

(1) Let $A_k = \begin{pmatrix} B_k & O \\ C_k & D_k \end{pmatrix}$ for each k.

$$\mathrm{rank}_{\mathbb{K}}(A) \geq \max\{\mathrm{rank}_{\mathbb{K}}(B_1; \ldots; B_p) + \mathrm{col\_rank}(D_1; \ldots; D_p),$$
$$\mathrm{rank}_{\mathbb{K}}(D_1; \ldots; D_p) + \mathrm{row\_rank}(B_1; \ldots; B_p)\}$$

(2) Let $A_k = (B_k, C_k)$ for each k and $1 \leq u < p$. If $C_k = O$, for all $1 \leq k \leq u$, then

$$\mathrm{rank}_{\mathbb{K}}(A) \geq \max\{\mathrm{rank}_{\mathbb{K}}(B_1; \ldots; B_u) + \mathrm{col\_rank}(C_{u+1}; \ldots; C_p),$$
$$\mathrm{rank}_{\mathbb{K}}(C_{u+1}; \ldots; C_p) + \dim\langle B_1, \ldots, B_u\rangle\}.$$

(3) Let $A_k = \begin{pmatrix} B_k \\ C_k \end{pmatrix}$ for each k and $1 \leq u < p$. If $C_k = O$, for all $1 \leq k \leq u$, then

$$\mathrm{rank}_{\mathbb{K}}(A) \geq \max\{\mathrm{rank}_{\mathbb{K}}(B_1; \ldots; B_u) + \mathrm{row\_rank}(C_{u+1}; \ldots; C_p),$$
$$\mathrm{rank}_{\mathbb{K}}(C_{u+1}; \ldots; C_p) + \dim\langle B_1, \ldots, B_u\rangle\}.$$

Proof By Theorem 5.17 (2), we have

$$\mathrm{rank}_{\mathbb{K}}(A) \geq \min_M \mathrm{rank}\left(\begin{pmatrix} B_1 \\ C_1 + D_1 M \end{pmatrix}; \ldots; \begin{pmatrix} B_p \\ C_p + D_p M \end{pmatrix}\right) + \mathrm{col\_rank}(D_1; \ldots; D_p)$$
$$\geq \mathrm{rank}_{\mathbb{K}}(B_1; \ldots; B_p) + \mathrm{col\_rank}(D_1; \ldots; D_p),$$

and by Theorem 5.17 (3),

$$\mathrm{rank}_{\mathbb{K}}(A) \geq \min_N \mathrm{rank}((C_1 + NB_1, D_1); \ldots; (C_p + NB_p, D_p)) + \mathrm{row\_rank}(B_1; \ldots; B_p)$$
$$\geq \mathrm{rank}_{\mathbb{K}}(D_1; \ldots; D_p) + \mathrm{row\_rank}(B_1; \ldots; B_p).$$

The second and third assertions are obtained from the first one by different slice direction or Theorem 5.17 (2), (1) and (3), (1), respectively.

Lemma 5.3 *Let $A = (A_1; A_2; \ldots; A_p) \in T_{\mathbb{K}}(m, n, p)$.*

(1) If $m < n$,

$$A_1 = (E_m, O_{m \times (n-m)}), \text{ and } A_k = (O_{m \times m}, B_k)$$

for $k \geq 2$, then $\mathrm{rank}_{\mathbb{K}}(A) \geq m + \mathrm{rank}_{\mathbb{K}}(B_2; \ldots; B_p)$.

(2) If $m \leq n$, $1 \leq t < m$,

$$A_1 = (E_m, O_{m \times (n-m)}), \ and \ A_k = \begin{pmatrix} O_{t \times t} & B_k \\ O_{(m-t) \times t} & O_{(m-t) \times (n-t)} \end{pmatrix}$$

for $k \geq 2$, then $\mathrm{rank}_{\mathbb{K}}(A) \geq m + \mathrm{rank}_{\mathbb{K}}(B_2; \ldots; B_p)$.

Proof (1) By Theorem 5.18 (2), we have

$$\mathrm{rank}_{\mathbb{K}}(A) \geq \mathrm{rank}_{\mathbb{K}}(B_2; \ldots; B_p) + \mathrm{col\_rank}(E_m) = m + \mathrm{rank}_{\mathbb{K}}(B_2; \ldots; B_p).$$

(2) If we apply Theorem 5.17 (2), then

$$\mathrm{rank}_{\mathbb{K}}(A) \geq \min_M \mathrm{rank}_{\mathbb{K}} \left(\begin{pmatrix} M \\ E_{m-t} \end{pmatrix}; \begin{pmatrix} B_2 \\ O_{(m-t) \times (n-t)} \end{pmatrix}; \ldots; \begin{pmatrix} B_p \\ O_{(m-t) \times (n-t)} \end{pmatrix} \right)$$
$$+ \mathrm{col\_rank}(E_t).$$

Next, we apply Theorem 5.17 (3). Then, we have

$$\mathrm{rank}_{\mathbb{K}}(A) \geq t + \min_N \mathrm{rank}_{\mathbb{K}}(N; B_2 \ldots; B_p) + \mathrm{row\_rank}(E_{m-t}; O; \ldots; O)$$
$$\geq m + \mathrm{rank}_{\mathbb{K}}(B_2 \ldots; B_p).$$

Theorem 5.19 *Suppose that $m \leq n$. Put $s = \lfloor n/m \rfloor \geq 1$, $c = n - ms \geq 0$, and $\ell_i = \lfloor (m + c)2^{-i} \rfloor$ for $1 \leq i \leq t$, and $p = s + t$. Let*

$$A_1 = (E_m, O_{m \times (n-m)}),$$
$$A_2 = (O_{m \times m}, E_m, O_{m \times (n-2m)}),$$
$$\vdots$$
$$A_s = (O_{m \times (s-1)m}, E_m, O_{m \times c}),$$
$$B_1 = \left(O_{m \times (n-\ell_1)}, \begin{pmatrix} E_{\ell_1} \\ O_{(m-\ell_1) \times \ell_1} \end{pmatrix} \right),$$
$$\vdots$$
$$B_t = \left(O_{m \times (n-\ell_t)}, \begin{pmatrix} E_{\ell_t} \\ O_{(m-\ell_t) \times \ell_t} \end{pmatrix} \right).$$

Then, $\mathrm{rank}_{\mathbb{K}}(A_1; \ldots; A_s; B_1; \ldots; B_t) = ms + \sum_{i=1}^t \ell_i$. In particular,

$$\mathrm{max.rank}_{\mathbb{K}}(m, n, s + t) \geq ms + \sum_{i=1}^t \ell_i.$$

Proof Clearly,

$$\text{rank}_{\mathbb{K}}(A_1; \ldots; A_s; B_1; \ldots; B_t) \leq \sum_{i=1}^{s} \text{rank}\, A_i + \sum_{i=1}^{t} \text{rank}\, B_i = ms + \sum_{i=1}^{t} \ell_i$$

by counting the nonzero elements. For the opposite inequality, we apply Lemma 5.3 (1) $(s-1)$ times repeatedly. Then, we have

$$\text{rank}_{\mathbb{K}}(A_1; \ldots; A_s; B_1; \ldots; B_t) \geq m(s-1) + \text{rank}_{\mathbb{K}}(A_s^{(1)}; B_1^{(1)}; \ldots; B_t^{(1)}),$$

where $A_s^{(1)} = (E_m, O_{m \times c})$ and $B_i^{(1)} = \left(O_{m \times (m+c-\ell_i)}, \begin{pmatrix} E_{\ell_i} \\ O_{(m-\ell_i) \times \ell_i} \end{pmatrix} \right)$. By

Lemma 5.3 (2), we see that

$$\text{rank}_{\mathbb{K}}(A_s^{(1)}; B_1^{(1)}; \ldots; B_t^{(1)}) \geq m + \text{rank}_{\mathbb{K}}(B_1^{(2)}; \ldots; B_t^{(2)}),$$

where $B_i^{(2)} = \left(O_{\ell_1 \times (\ell_1 - \ell_i)}, \begin{pmatrix} E_{\ell_i} \\ O_{(\ell_1 - \ell_i) \times \ell_i} \end{pmatrix} \right)$ for $i \geq 1$. Again, by Lemma 5.3 (2), we

see that

$$\text{rank}_{\mathbb{K}}(B_1^{(2)}; \ldots; B_t^{(2)}) \geq \ell_1 + \text{rank}_{\mathbb{K}}(B_2^{(3)}; \ldots; B_t^{(3)}),$$

where

$$B_i^{(3)} = \left(O_{\ell_2 \times (\ell_2 - \ell_i)}, \begin{pmatrix} E_{\ell_i} \\ O_{(\ell_2 - \ell_i) \times \ell_i} \end{pmatrix} \right)$$

for $i \geq 2$. Therefore, we have

$$\text{rank}_{\mathbb{K}}(A_1; \ldots; A_s; B_1; \ldots; B_t) \geq ms + \ell_1 + \text{rank}_{\mathbb{K}}(B_2^{(3)}; \ldots; B_t^{(3)}).$$

Hence, inductively, we get the assertion.

Corollary 5.2

$$\text{max.rank}_{\mathbb{K}}(2^k, 2^k, k+1) \geq 2^{k+1} - 1.$$

Proof In Theorem 5.19, we consider the case where $m = n = 2^k$ and $t = k$. Then, the tensor $(A_1; B_1; \ldots; B_k)$ has rank $\sum_{j=0}^{k} 2^{k-j} = 2^{k+1} - 1$.

Chapter 6
Typical Ranks

Let m, n, and p be positive integers. In this chapter, we discuss the typical ranks of $m \times n \times p$ tensors over $\mathbb{R}$ and the generic rank of $m \times n \times p$ tensors over $\mathbb{C}$. For the readers' convenience, some basic facts of algebraic geometry are included.

6.1 Introduction

In this chapter, we consider typical ranks of 3-tensors over $\mathbb{R}$ and the generic rank of 3-tensors over $\mathbb{C}$ of fixed size.

Consider the matrix case, i.e., the 2-tensor case. Let M be an $m \times n$ matrix with $m \leq n$. Then, almost always $\text{rank}\, M = m$. In fact, $\text{rank}\, M$ is always less than or equal to m and if one of the maximal minors of M does not vanish, then $\text{rank}\, M = m$. Thus the set of $m \times n$ matrices whose rank is not m is the intersection of the zero loci of the $\binom{n}{m}$ polynomials of the entry. Thus, the set of $m \times n$ matrices whose rank is not m is a thin set and "almost always" an $m \times n$ matrix has rank m.

The phenomenon is quite different in the case of 3 or higher dimensional tensors. For example, consider a $2 \times 2 \times 2$ tensor $T = (T_1; T_2)$. As noted above, T_1 is almost always nonsingular. Thus, by multiplying T_1^{-1}, we consider a tensor $S = (E_2; A)$. Then, as shown in Chap. 2 $\text{rank}\, S = 2$ if and only if A is diagonalizable. If we are working over $\mathbb{C}$, an algebraically closed field, then the condition that the characteristic polynomial of A has no multiple roots is sufficient for A to be diagonalizable. A polynomial has multiple root if and only if its discriminant is 0, thus, the set of A that is not diagonalizable is a thin set. Therefore, if the base field is $\mathbb{C}$, then a $2 \times 2 \times 2$ tensor almost always has rank 2.

Now suppose that the base field is $\mathbb{R}$. A matrix close to $\begin{pmatrix} 0 & -1 \\ 1 & 0 \end{pmatrix}$ has imaginary eigenvalues and is thus not diagonalizable over $\mathbb{R}$. Therefore, a tensor close to $\left(E_2; \begin{pmatrix} 0 & -1 \\ 1 & 0 \end{pmatrix} \right)$ has rank more than 2 (in fact 3, see Sumi et al. 2009). Thus, there is a Euclidean open subset of $\mathbb{R}^{2 \times 2 \times 2}$ whose elements all have rank 3. On the other

© The Author(s) 2016
T. Sakata et al., *Algebraic and Computational Aspects of Real Tensor Ranks*,
JSS Research Series in Statistics, DOI 10.1007/978-4-431-55459-2_6

hand, a tensor close to $\left(E_2; \begin{pmatrix} 1 & 0 \\ 0 & 2 \end{pmatrix} \right)$ has rank 2. Thus, there is also a Euclidean open subset of $\mathbb{R}^{2\times2\times2}$ whose elements all have rank 2. Thus, if one choses an element T of $S^7 \subset \mathbb{R}^{2\times2\times2}$ randomly, the probabilities $\mathrm{rank} T = 2$ and $\mathrm{rank} T = 3$ are both positive. Note that the rank of a tensor is invariant under the multiplication of a nonzero scalar.

Let m, n, and p be integers greater than 1. If the probability that $\mathrm{rank} T = r$ is positive, where T is a randomly chosen element of $S^{mnp-1} \subset \mathbb{R}^{m\times n\times p}$, r is called a typical rank of $m \times n \times p$ tensors over $\mathbb{R}$. We consider in the following sections, the typical ranks over $\mathbb{R}$. We also show that if one considers a counterpart of the typical rank over $\mathbb{C}$, then there is only one such value for any m, n, and p and it coincides with the minimal typical rank of $m \times n \times p$ tensors over $\mathbb{R}$. We call this the generic rank of $m \times n \times p$ tensors over $\mathbb{C}$.

6.2 Generic Rank Over $\mathbb{C}$ and Typical Rank Over $\mathbb{R}$

In this section, we summarize some basic facts on algebraic geometry. In addition, we define generic rank over $\mathbb{C}$ and typical rank over $\mathbb{R}$. The basic references on algebraic geometry are Harris 1992 and Cox et al. 1992.

Let $\mathbb{K}$ be an infinite field throughout this chapter. Let n be a positive integer and $X_1, ..., X_n$ be indeterminates. For a subset S of $\mathbb{K}[X_1, ..., X_n]$, the polynomial ring with n variables, we set

$$\mathbb{V}(S) := \{(a_1, ..., a_n) \in \mathbb{K}^n \mid f(a_1, ..., a_n) = 0 \text{ for any } f \in S\}$$

and for a subset W of $\mathbb{K}^n$, we define

$$\mathbb{I}(W) := \{f \in \mathbb{K}[X_1, ..., X_n] \mid f(a) = 0 \text{ for any } a \in W\}.$$

Note that $\mathbb{I}(W)$ is an ideal of $\mathbb{K}[X_1, ..., X_n]$. If $S = \{f_1, ..., f_t\}$, we write $\mathbb{V}(S)$ as $\mathbb{V}(f_1, ..., f_t)$ for simplicity.

The following results are easily verified. (for (5), see, e.g., Cox et al. 1992, Chap. 1 Sect. 1 Proposition 5).

Lemma 6.1 *(1) If $S_1 \subset S_2$, then $\mathbb{V}(S_1) \supset \mathbb{V}(S_2)$.*
(2) If $W_1 \subset W_2$, then $\mathbb{I}(W_1) \supset \mathbb{I}(W_2)$.
(3) $\mathbb{I}(\mathbb{V}(S)) \supset S$, $\mathbb{V}(\mathbb{I}(W)) \supset W$.
(4) $\mathbb{V}(\mathbb{I}(\mathbb{V}(S))) = \mathbb{V}(S)$, $\mathbb{I}(\mathbb{V}(\mathbb{I}(W))) = \mathbb{I}(W)$.
(5) $\mathbb{I}(\mathbb{K}^n) = \langle 0 \rangle$.

Definition 6.1 An affine algebraic variety in $\mathbb{K}^n$ is a subset of $\mathbb{K}^n$ of the form $\mathbb{V}(S)$ for some $S \subset \mathbb{K}[X_1, ..., X_n]$.

We note the following basic fact.

Lemma 6.2 *(1)* $\emptyset$ *and* $\mathbb{K}^n$ *are affine algebraic varieties.*
(2) If V_1 *and* V_2 *are affine algebraic varieties, then so is* $V_1 \cup V_2$.
(3) If $\{V_\lambda\}_{\lambda \in \Lambda}$ *is a family of affine algebraic varieties, then* $\bigcap_{\lambda \in \Lambda} V_\lambda$ *is an affine algebraic variety.*

Proof (1) $\mathbb{V}(1) = \emptyset$ and $\mathbb{V}(0) = \mathbb{K}^n$.
(2) Suppose that $V_i = \mathbb{V}(S_i)$ for $i = 1, 2$. Set $S = \{fg \mid f \in S_1, g \in S_2\}$. Then, $V_1 \cup V_2 = \mathbb{V}(S)$.
(3) Suppose that $V_\lambda = \mathbb{V}(S_\lambda)$ for $\lambda \in \Lambda$. Then $\bigcap_{\lambda \in \Lambda} V_\lambda = \mathbb{V}(\bigcup_{\lambda \in \Lambda} S_\lambda)$.

By Lemma 6.2, we see that there is a topology on $\mathbb{K}^n$ whose closed sets are affine algebraic varieties.

Definition 6.2 The Zariski topology on $\mathbb{K}^n$ is the topology on $\mathbb{K}^n$ whose closed sets are affine algebraic varieties. For any affine algebraic variety V in $\mathbb{K}^n$, we introduce the topology, which is also called the Zariski topology on V, as the induced topology on V from the Zariski topology of $\mathbb{K}^n$.

In the remainder of this chapter, when we use terms concerning topology, they are based on the Zariski topology, except for the case explicitly referring to other topologies.

Lemma 6.3 *Let* V *be an affine algebraic variety in* $\mathbb{K}^n$, *O be a subset open in* V *and* $a \in O$. *Then, there exists* $f \in \mathbb{K}[X_1, \ldots, X_n]$ *such that*

$$a \in V \backslash \mathbb{V}(f) \subset O.$$

Proof Take a subset S of $\mathbb{K}[X_1, \ldots, X_n]$ such that $O = V \backslash \mathbb{V}(S)$. Since $a \in O$, $a \notin \mathbb{V}(S)$. Thus there exists $f \in S$ such that $f(a) \neq 0$. Then, $a \in V \backslash \mathbb{V}(f) \subset O$.

Remark 6.1 Suppose that $\mathbb{K} = \mathbb{R}$ or $\mathbb{C}$. Then, a Zariski closed set is closed in the Euclidean topology.

Remark 6.2 Suppose that $\mathbb{K} = \mathbb{R}$ or $\mathbb{C}$ and V is an affine algebraic variety in $\mathbb{K}^n$ such that $V \neq \mathbb{K}^n$. If $a \in \mathbb{K}^n$ moves randomly according to a distribution whose probability density function is positive anywhere, the probability that $a \in V$ is 0.

Example 6.1 $\mathbb{K}^{n \times n}$, the set of $n \times n$ matrices with entries in $\mathbb{K}$ can be identified with $\mathbb{K}^{n^2}$. Since $\det A$ is a polynomial of entries of A, we see that

$$\mathrm{SL}(n, \mathbb{K}) = \{A \in \mathbb{K}^{n \times n} \mid \det A - 1 = 0\}$$

is an affine algebraic variety in $\mathbb{K}^{n \times n}$.

Example 6.2 $GL(n, \mathbb{K}) = \{A \in \mathbb{K}^{n \times n} \mid \det A \neq 0\}$ cannot be treated as an affine algebraic variety directly. However, we can identify $GL(n, \mathbb{K})$ with the affine algebraic variety

$$\{(A, b) \in \mathbb{K}^{n \times n} \times \mathbb{K} \mid b(\det A) - 1 = 0\}$$

in $\mathbb{K}^{n^2+1}$.

Definition 6.3 An affine algebraic variety V is said to be irreducible if $V = V_1 \cup V_2$, where V_1 and V_2 are affine algebraic varieties, then $V = V_1$ or $V = V_2$.

The following lemma is easily verified:

Lemma 6.4 *Let V be an affine algebraic variety. Then, the following conditions are equivalent:*

(1) V is irreducible.
(2) For any two nonempty open sets O_1 and O_2 of V, $O_1 \cap O_2 \neq \emptyset$.
(3) Any nonempty open set of V is dense in V.

Lemma 6.5 $\mathbb{K}^n$ *is an irreducible affine algebraic variety.*

Proof Suppose that V_1 and V_2 are affine algebraic varieties in $\mathbb{K}^n$ such that $V_i \subsetneq \mathbb{K}^n$ for $i = 1, 2$. Set $V_i = \mathbb{V}(S_i)$ for $i = 1, 2$. Since $V_i \subsetneq \mathbb{K}^n$, we see that $S_i \setminus \{0\} \neq \emptyset$ for $i = 1, 2$. Take $0 \neq f_i \in S_i$ for $i = 1, 2$. Then, $f_1 f_2 \neq 0$. Thus,

$$V_1 \cup V_2 \subset \mathbb{V}(f_1 f_2) \subsetneq \mathbb{K}^n$$

since $\mathbb{K}$ is an infinite field.

Definition 6.4 Let V be an affine algebraic variety in $\mathbb{K}^n$.

(1) A regular function or a polynomial function on V is a map $F \colon V \to \mathbb{K}$ such that there exists a polynomial $f \in \mathbb{K}[X_1, \ldots, X_n]$ such that $F(a) = f(a)$ for any $a \in V$.
(2) Let W be an affine algebraic variety in $\mathbb{K}^m$. A regular map or a polynomial map from V to W is a map $F \colon V \to W$ such that there exist regular functions $F_1, \ldots, F_m$ on V with $F(a) = (F_1(a), \ldots, F_m(a))$ for any $a \in V$.

Here we state a basic but important fact about regular maps.

Lemma 6.6 *Let V be an affine algebraic variety in $\mathbb{K}^n$ and W be an affine algebraic variety in $\mathbb{K}^m$, $X_1, \ldots, X_n, Y_1, \ldots, Y_m$ indeterminates. We use $\mathbb{K}[X_1, \ldots, X_n]$ to explain the facts concerning affine algebraic varieties in $\mathbb{K}^n$ and $\mathbb{K}[Y_1, \ldots, Y_m]$ in $\mathbb{K}^m$.*
Suppose that $f_1, \ldots, f_m \in \mathbb{K}[X_1, \ldots, X_n]$ and let $F \colon \mathbb{K}^n \to \mathbb{K}^m$ be the regular map defined by $f_1, \ldots, f_m$, i.e., $F(a) = (f_1(a), \ldots, f_m(a))$ for $a \in \mathbb{K}^n$. Let $\tilde{F} \colon \mathbb{K}[Y_1, \ldots, Y_m] \to \mathbb{K}[X_1, \ldots, X_n]$ be the $\mathbb{K}$-algebra homomorphism mapping Y_i to f_i for $1 \leq i \leq m$. Then

(1) $F(V) \subset W$ *if and only if* $\widetilde{F}(\mathbb{I}(W)) \subset \mathbb{I}(V)$ *and*
(2) $\mathbb{V}(\widetilde{F}^{-1}(\mathbb{I}(V)))$ *is the closure of* $F(V)$.

Proof First, note that for $g \in \mathbb{K}[Y_1, \ldots, Y_m]$, $\widetilde{F}(g) = g(f_1(X_1, \ldots, X_n), \ldots, f_m(X_1, \ldots, X_n))$. Thus, $\widetilde{F}(g)(a) = g(f_1(a), \ldots, f_m(a)) = g(F(a))$ for any $a \in \mathbb{K}^n$.

(1) First, suppose that $F(V) \subset W$ and let g be an arbitrary element of $\mathbb{I}(W)$. Then for any $a = (a_1, \ldots, a_n) \in V$, $F(a) = (f_1(a), \ldots, f_m(a)) \in W$. Therefore, $\widetilde{F}(g)(a) = g(f_1(a), \ldots, f_m(a)) = 0$ since $g \in \mathbb{I}(W)$. Since a is an arbitrary element of V, we see that $\widetilde{F}(g) \in \mathbb{I}(V)$. Thus, we see that $\widetilde{F}(\mathbb{I}(W)) \subset \mathbb{I}(V)$.

Next, assume that $\widetilde{F}(\mathbb{I}(W)) \subset \mathbb{I}(V)$ and let a be an arbitrary element of V. For any $g \in \mathbb{I}(W)$, $\widetilde{F}(g) \in \mathbb{I}(V)$ by assumption. Thus, $\widetilde{F}(g)(a) = 0$. Since $\widetilde{F}(g)(a) = g(f_1(a), \ldots, f_m(a)) = g(F(a))$, we see that

$$g(F(a)) = 0 \quad \text{for any } g \in \mathbb{I}(W).$$

Thus, $F(a) \in \mathbb{V}(\mathbb{I}(W)) = W$ by Lemma 6.1.

(2) First, we show that $\mathbb{V}(\widetilde{F}^{-1}(\mathbb{I}(V))) \supset F(V)$. Let b be an arbitrary element of $F(V)$. Take $a \in V$ such that $b = F(a)$. Then for any $g \in \widetilde{F}^{-1}(\mathbb{I}(V))$, $g(F(a)) = \widetilde{F}(g)(a) = 0$. Thus, $b = F(a) \in \mathbb{V}(\widetilde{F}^{-1}(\mathbb{I}(V)))$. Thus, we see that $\mathbb{V}(\widetilde{F}^{-1}(\mathbb{I}(V)))$ is an affine algebraic variety containing $F(V)$.

Now let W' be an arbitrary affine algebraic variety in $\mathbb{K}^m$ containing $F(V)$. Then by (1), we see that $\widetilde{F}(\mathbb{I}(W')) \subset \mathbb{I}(V)$. Thus, we see that $\mathbb{I}(W') \subset \widetilde{F}^{-1}(\mathbb{I}(V))$ and $W' = \mathbb{V}(\mathbb{I}(W')) \supset \mathbb{V}(\widetilde{F}^{-1}(\mathbb{I}(V)))$ by Lemma 6.1. Thus, $\mathbb{V}(\widetilde{F}^{-1}(\mathbb{I}(V)))$ is the smallest affine algebraic variety containing $F(V)$.

Corollary 6.1 *In the notation of Lemma 6.6, $\mathrm{Im}\, F$ is dense in $\mathbb{K}^m$ if and only if $\widetilde{F}$ is injective.*

Proof By Lemma 6.6 (2) and Lemma 6.1 (5), we see that the closure of $\mathrm{Im}\, F$ is $\mathbb{V}(\widetilde{F}^{-1}(\langle 0 \rangle)) = \mathbb{V}(\ker \widetilde{F})$. If $\ker \widetilde{F} = \langle 0 \rangle$, then $\mathbb{V}(\ker \widetilde{F}) = \mathbb{K}^m$. If $\ker \widetilde{F} \neq \langle 0 \rangle$, then there exists $g \in \ker \widetilde{F}$ with $g \neq 0$. Therefore,

$$\mathbb{V}(\ker \widetilde{F}) \subset \mathbb{V}(g) \subsetneq \mathbb{K}^m$$

since $\mathbb{K}$ is an infinite field.

Let m, n, and p be positive integers. Then, $\mathbb{K}^{m \times n \times p}$, the set of $m \times n \times p$ tensors with entries in $\mathbb{K}$ is naturally identified with $\mathbb{K}^{mnp}$. Consider the following regular maps:

$$\Phi_1^{\mathbb{K}}: \mathbb{K}^m \times \mathbb{K}^n \times \mathbb{K}^p \ni (a_1, \ldots, a_m, b_1, \ldots, b_n, c_1, \ldots, c_p) \mapsto (a_i b_j c_k) \in \mathbb{K}^{m \times n \times p}$$

and

$$\Phi_r^{\mathbb{K}}: (\mathbb{K}^m \times \mathbb{K}^n \times \mathbb{K}^p)^r \ni (x_1, \ldots, x_r) \mapsto \Phi_1^{\mathbb{K}}(x_1) + \cdots + \Phi_1^{\mathbb{K}}(x_r) \in \mathbb{K}^{m \times n \times p},$$

where r is a positive integer. Then, the image of $\Phi_r^{\mathbb{K}}$ is the set of tensors whose rank is less than or equal to r. We write Φ_r for $\Phi_r^{\mathbb{K}}$ if $\mathbb{K}$ is clear from the context.

Now assume that $\mathbb{K} = \mathbb{R}$ or $\mathbb{C}$. If $\mathrm{Im}\Phi_r$ is not dense in $\mathbb{K}^{m \times n \times p}$, then the closure of $\mathrm{Im}\Phi_r$ is a proper closed subset of $\mathbb{K}^{m \times n \times p}$. Therefore, if an $m \times n \times p$ tensor T moves randomly according to a distribution whose probability density function is positive anywhere, the probability that $\mathrm{rank} T \leq r$ is 0 by Remark 6.2

Now consider the "generic rank" of tensors over $\mathbb{C}$ of fixed size. First, we cite the following fact Northcott 1980, Chap. 3 Theorem 33, which is usually proved using the theorem of Chevalley (see Harris 1992, Theorem 3.16).

Theorem 6.1 *Let $\mathbb{K}$ be an algebraically closed field, V be an irreducible affine algebraic variety over $\mathbb{K}$, W be an affine algebraic variety over $\mathbb{K}$, and $F : V \to W$ be a regular map. If $\mathrm{Im} F$ is dense in W, then $\mathrm{Im} F$ contains a nonempty open subset of W.*

Theorem 6.2 *Let m, n, and p be positive integers. Suppose that r is the minimum integer such that $\mathrm{Im}\Phi_r^{\mathbb{C}}$ is dense in $\mathbb{C}^{m \times n \times p}$. Then, there exists a dense open subset $\mathcal{O}$ of $\mathbb{C}^{m \times n \times p}$ such that for any $T \in \mathcal{O}$, $\mathrm{rank} T = r$. In particular, if $T \in \mathbb{C}^{m \times n \times p}$ moves randomly according to a distribution whose probability density function is positive anywhere, then the probability that $\mathrm{rank} T = r$ is 1.*

Proof By Theorem 6.1, we see that there exists a nonempty open subset $\mathcal{U}$ of $\mathbb{C}^{m \times n \times p}$ that is contained in $\mathrm{Im}\Phi_r^{\mathbb{C}}$. Since the closure of $\mathrm{Im}\Phi_{r-1}^{\mathbb{C}}$ is a proper subset of $\mathbb{C}^{m \times n \times p}$ by assumption, we see that the complement of the closure of $\mathrm{Im}\Phi_{r-1}^{\mathbb{C}}$ is a nonempty open subset of $\mathbb{C}^{m \times n \times p}$. Since $\mathbb{C}^{m \times n \times p}$ is irreducible by Lemma 6.5, $\mathcal{O} := \mathcal{U} \backslash ($the closure of $\mathrm{Im}\Phi_{r-1}^{\mathbb{C}})$ is a dense open subset of $\mathbb{C}^{m \times n \times p}$.

If $T \in \mathcal{O}$, then $T \in \mathrm{Im}\Phi_r^{\mathbb{C}}$ and $T \notin \mathrm{Im}\Phi_{r-1}^{\mathbb{C}}$. Therefore, $\mathrm{rank} T = r$. The last statement follows from Remark 6.2.

Definition 6.5 We call r in Theorem 6.2 the generic rank of $m \times n \times p$ tensors over $\mathbb{C}$ and denote $r = \mathrm{grank}_{\mathbb{C}}(m, n, p)$.

Next, we consider "typical ranks" of tensors over $\mathbb{R}$. First, we cite the following basic fact on commutative algebra (see, e.g., Matsumura 1989, Appendix A Formulas 5 and 8).

Theorem 6.3 *Let $\mathbb{K}$ be a field, R be a commutative ring that contains $\mathbb{K}$ as a subring, $X_1, ..., X_n$ and $Y_1, ..., Y_m$ be indeterminates and $f_1,...,f_m \in \mathbb{K}[X_1, ..., X_n]$. Further, let $\varphi_{\mathbb{K}}$ (resp. φ_R) be the $\mathbb{K}$-algebra (resp. R-algebra) homomorphism $\mathbb{K}[Y_1, ..., Y_m] \to \mathbb{K}[X_1, ..., X_n]$ (resp. $R[Y_1, ..., Y_m] \to R[X_1, ..., X_n]$) sending Y_i to f_i for $1 \leq i \leq m$. Then, $\varphi_{\mathbb{K}}$ is injective if and only if φ_R is injective.*

Corollary 6.2 *Let r be a positive integer. Then $\widetilde{\Phi_r^{\mathbb{R}}}$ is injective if and only if $\widetilde{\Phi_r^{\mathbb{C}}}$ is injective. In particular, $\mathrm{Im}\Phi_r^{\mathbb{R}}$ is dense in $\mathbb{R}^{m \times n \times p}$ if and only if $\mathrm{Im}\Phi_r^{\mathbb{C}}$ is dense in $\mathbb{C}^{m \times n \times p}$.*

Proof The first assertion follows from Theorem 6.3. The last assertion follows from the first one and Corollary 6.1.

Here we state the following basic fact without proof:

Lemma 6.7 *Let m and n be positive integers and $f_1, \ldots, f_m \in \mathbb{C}[X_1, \ldots, X_n]$. Set*

$$F : \mathbb{C}^n \ni a \mapsto (f_1(a), \ldots, f_m(a)) \in \mathbb{C}^m.$$

If $\operatorname{Im} F$ *has an interior point with respect to the Euclidean topology, then the Jacobian matrix*

$$\frac{\partial(f_1, \ldots, f_m)}{\partial(X_1, \ldots, X_n)}$$

has rank m.

Now we state the following:

Definition 6.6 Suppose that $T \in \mathbb{R}^{m \times n \times p}$ moves randomly according to a distribution whose probability density function is positive anywhere. If the probability that $\operatorname{rank} T = r$ is positive, then we say that r is a typical rank of $m \times n \times p$ tensors over $\mathbb{R}$. The set of typical ranks of $m \times n \times p$ tensors over $\mathbb{R}$ is denoted as $\operatorname{trank}_{\mathbb{R}}(m, n, p)$.

Consider the typical ranks of tensors over $\mathbb{R}$ with two slices. First, we recall our previous result, Miyazaki et al. (2009).

Theorem 6.4 *Let $\mathbb{K}$ be an infinite field. Suppose that $1 \leq n < p$. Then, there exist rational maps*

$$\varphi_1^{n \times p} : \mathbb{K}^{n \times p \times 2} \dashrightarrow \operatorname{GL}(n, \mathbb{K})$$
$$\varphi_2^{n \times p} : \mathbb{K}^{n \times p \times 2} \dashrightarrow \operatorname{GL}(p, \mathbb{K})$$

such that

$$\varphi_1^{n \times p}(T) T \varphi_2^{n \times p}(T) = ((E_n, O); (O, E_n)),$$

for any $T \in \operatorname{dom} \varphi_1^{n \times p} \cap \operatorname{dom} \varphi_2^{n \times p}$,

$$((E_n, O); (O, E_n)) \in \operatorname{dom} \varphi_1^{n \times p} \cap \operatorname{dom} \varphi_2^{n \times p},$$

$$\varphi_1^{n \times p}((E_n, O); (O, E_n)) = E_n,$$

and

$$\varphi_2^{n \times p}((E_n, O); (O, E_n)) = E_p.$$

(See Sect. 6.3 for the definition of a rational map.)

Since tensors with two slices are classified (Theorem 5.1), we see the following fact by Theorem 5.3.

Theorem 6.5 (ten Berge and Kiers 1999) *Let* $2 \leq m < n$. *The typical rank of* $\mathbb{R}^{m \times m \times 2}$ *is* $\{m, m+1\}$ *and the typical rank of* $\mathbb{R}^{m \times n \times 2}$ *is* $\{\min(n, 2m)\}$.

Proof The set $S(m)$ consisting of all $(A; B) \in \mathbb{R}^{m \times m \times 2}$ such that $\det(A) \neq 0$ and $A^{-1}B$ has distinct nonzero real eigenvalues contains a nonempty Euclidean open set and consists of tensors with rank m. The set $S(m+1)$ consisting of all $(A; B) \in \mathbb{R}^{m \times m \times 2}$ such that $\det(A) \neq 0$ and $A^{-1}B$ has distinct imaginary eigenvalues contains a nonempty Euclidean open set and consists of tensors with rank $m+1$. It is easy to see that $\mathbb{R}^{m \times m \times 2} \setminus (S(m) \cup S(m+1))$ does not contain a Euclidean open set. Therefore, $\mathrm{trank}_{\mathbb{R}}(m, m, 2) = \{m, m+1\}$.

Next, consider $\mathrm{trank}_{\mathbb{R}}(m, n, 2)$. Set $\mathcal{O} = \mathrm{dom}\varphi_1^{m \times n} \cap \mathrm{dom}\varphi_2^{m \times n}$ in the notation of Theorem 6.4. Then, we see that $\mathcal{O}$ is a dense open subset of $\mathbb{R}^{m \times n \times 2}$ (see Remark 6.3 and Definition 6.8 of Sect. 6.3) such that if $T \in \mathcal{O}$, then T is $\mathrm{GL}(m) \times \mathrm{GL}(n)$-equivalent to $((E_m, O); (O, E_m))$. Since $\mathrm{rank}((E_m, O); (O, E_m)) = \min(n, 2m)$, we see that $\mathrm{trank}_{\mathbb{R}}(m, n, 2) = \min(n, 2m)$.

Note that there exist integers m, n, and p such that there are multiple typical ranks of $m \times n \times p$ tensors over $\mathbb{R}$.

Theorem 6.6 *Let* m, n, *and* p *be positive integers. Set* $r_0 = \mathrm{grank}_{\mathbb{C}}(m, n, p)$. *Then,* r_0 *is the minimal typical rank of* $m \times n \times p$ *tensors over* $\mathbb{R}$.

Proof Suppose that $r < r_0$. Then, $\mathrm{Im}\Phi_r^{\mathbb{C}}$ is not dense in $\mathbb{C}^{m \times n \times p}$. Therefore, $\mathrm{Im}\Phi_r^{\mathbb{R}}$ is not dense in $\mathbb{R}^{m \times n \times p}$ by Corollary 6.2. Thus, the closure of $\mathrm{Im}\Phi_r^{\mathbb{R}}$ is a proper closed subset of $\mathbb{R}^{m \times n \times p}$. Thus, if $T \in \mathbb{R}^{m \times n \times p}$ moves randomly according to a distribution whose probability density function is positive anywhere, the probability that $\mathrm{rank}\,T \leq r$ is 0 by Remark 6.2.

Next, by Lemma 6.7, we see that the Jacobian matrix of $\Phi_{r_0}^{\mathbb{C}}$ is full row rank. Since all the coefficients of $\Phi_{r_0}^{\mathbb{C}}$ are real numbers, we see that the Jacobian matrix of $\Phi_{r_0}^{\mathbb{R}}$ is also full row rank. Thus, there exists $a \in (\mathbb{R}^m \times \mathbb{R}^n \times \mathbb{R}^p)^{r_0}$ such that the Jacobian matrix of $\Phi_{r_0}^{\mathbb{R}}$ is full row rank at a. Then we see that $\Phi_{r_0}^{\mathbb{R}}(a)$ is an interior point of $\mathrm{Im}\Phi_{r_0}^{\mathbb{R}}$ with respect to the Euclidean topology. Thus if $T \in \mathbb{R}^{m \times n \times p}$ moves randomly according to a distribution whose probability density function is positive anywhere, the probability that $T \in \mathrm{Im}\Phi_{r_0}^{\mathbb{R}}$ is positive. Since probability that $\mathrm{rank}\,T < r_0$ is 0, we see that the probability that $\mathrm{rank}\,T = r_0$ is positive.

Therefore, $r_0 = \min \mathrm{trank}_{\mathbb{R}}(m, n, p)$.

The following fact is known.

Theorem 6.7 (Friedland 2012, Theorem 7.1) *The space* $\mathbb{R}^{m \times n \times p}$ *contains a finite number of Euclidean open connected disjoint sets* $O_1, \ldots, O_M$ *satisfying the following properties:*

(1) $\mathbb{R}^{m \times n \times p} \setminus \cup_{i=1}^M O_i$ *is a Euclidean closed set of* $\mathbb{R}^{m \times n \times p}$ *of dimension less than* mnp.
(2) *Each* $T \in O_i$ *has rank* r_i *for* $i = 1, \ldots, M$.
(3) *The number* $\min(r_1, \ldots, r_M)$ *is equal to* $\mathrm{grank}_{\mathbb{C}}(m, n, p)$.

(4) $max(r_1, \ldots, r_M)$ is the minimal integer t such that the Euclidean closure of $\operatorname{Im}\Phi_t^{\mathbb{R}}$ is equal to $\mathbb{R}^{m \times n \times p}$.

(5) For each integer $r \in [min(r_1, \ldots, r_M), max(r_1, \ldots, r_M)]$, there exists $r_i = r$ for some integer $i \in [1, M]$.

6.3 Rational Functions and Rational Maps

In this section, we summarize the definition and basic facts of rational functions and rational maps, which are basic notions in algebraic geometry.

Let V be an irreducible affine algebraic variety in $\mathbb{K}^n$. Consider the following set.

$$\{(f, g) \mid f, g \in \mathbb{K}[X_1, \ldots, X_n], g \notin \mathbb{I}(V)\}$$

We define a binary relation $\sim$ on this set by

$$(f_1, g_1) \sim (f_2, g_2) \overset{\text{def}}{\Longleftrightarrow} f_1(x)g_2(x) = f_2(x)g_1(x) \quad \text{for any } x \in V.$$

Lemma 6.8 $\sim$ *is an equivalence relation.*

Proof Reflexivity and symmetricity are trivial. Suppose that $(f_1, g_1) \sim (f_2, g_2)$ and $(f_2, g_2) \sim (f_3, g_3)$. Then

$$f_1(x)g_3(x)g_2(x) = f_2(x)g_1(x)g_3(x) = f_3(x)g_1(x)g_2(x)$$

for any $x \in V$. Thus

$$f_1(x)g_3(x) = f_3(x)g_1(x) \quad \text{for any } x \in V \text{ with } g_2(x) \neq 0.$$

Since V is irreducible and $g_2 \notin \mathbb{I}(V)$, the set $\{x \in V \mid g_2(x) \neq 0\}$ is a dense open subset of V. Because $\{x \in V \mid f_1(x)g_3(x) = f_3(x)g_1(x)\}$ is a closed subset of V containing the above dense open subset of V, we see that $f_1(x)g_3(x) = f_3(x)g_1(x)$ for any $x \in V$. That is, $(f_1, g_1) \sim (f_3, g_3)$. $\square$

Definition 6.7 A rational function on an irreducible affine algebraic variety V is an equivalence class with respect to $\sim$. Let φ be a rational function on V and (f, g) a representative of φ. We define the domain of φ, written $\operatorname{dom}\varphi$ by

$$\operatorname{dom}\varphi := \bigcup_{(f', g') \sim (f, g)} \{x \in V \mid g'(x) \neq 0\}.$$

For $x \in \operatorname{dom}\varphi$, we can naturally define the value of φ at x: take a representative (f', g') of φ such that $g'(x) \neq 0$ and we define the value of φ at x as $f'(x)/g'(x)$. We denote the value of φ at x by $\varphi(x)$.

It is easily verified that the value of φ at x defined above is independent of the choice of the representative of φ.

Remark 6.3 Let φ be a rational function on V. Then, $\mathrm{dom}\varphi$ is a dense open subset of V.

Lemma 6.9 *Let φ_1 and φ_2 be rational functions on an irreducible affine algebraic variety V. Suppose that there exists a dense open subset O of V such that*

$$\varphi_1(x) = \varphi_2(x) \quad \text{for any } x \in \mathrm{dom}\varphi_1 \cap \mathrm{dom}\varphi_2 \cap O.$$

Then, $\varphi_1 = \varphi_2$.

Proof Take representatives (f_1, g_1) and (f_2, g_2) of φ_1 and φ_2, respectively. Then

$$f_1(x)g_2(x) = f_2(x)g_1(x) \quad \text{for any } x \in \mathrm{dom}\varphi_1 \cap \mathrm{dom}\varphi_2 \cap O$$

by assumption. Since $\mathrm{dom}\varphi_1 \cap \mathrm{dom}\varphi_2 \cap O$ is a dense open subset of V, we see that

$$f_1(x)g_2(x) = f_2(x)g_1(x) \quad \text{for any } x \in V.$$

Thus, $(f_1, g_1) \sim (f_2, g_2)$ and $\varphi_1 = \varphi_2$.

Definition 6.8 Let V be an irreducible affine algebraic variety in $\mathbb{K}^n$ and W be an affine algebraic variety in $\mathbb{K}^m$. A rational map φ from V to W is an m-tuple of rational functions $(\varphi_1, \ldots, \varphi_m)$ on V such that $(\varphi_1(x), \ldots, \varphi_m(x)) \in W$ for any $x \in \bigcap_{i=1}^{m} \mathrm{dom}\varphi_i$. We define $\mathrm{dom}\varphi := \bigcap_{i=1}^{m} \mathrm{dom}\varphi_i$ and for $x \in \mathrm{dom}\varphi$, we define the image of x by φ as $(\varphi_1(x), \ldots, \varphi_m(x))$. We denote by $\varphi \colon V - - \to W$ that φ is a rational map from V to W.

Remark 6.4 A regular map from an irreducible variety is a rational map.

Next, we consider the composition of rational maps. First, we note the following fact whose proof is easy.

Lemma 6.10 *Let $X_1, \ldots, X_n$ and $Y_1, \ldots, Y_m$ be indeterminates, $h \in \mathbb{K}[Y_1, \ldots, Y_m]$, and $f_1, \ldots, f_m, g_1, \ldots, g_m \in \mathbb{K}[X_1, \ldots, X_n]$. Suppose that $g_1 \ldots g_m \neq 0$. Then, there exists a positive integer d such that*

$$(g_1 \ldots g_m)^d h(f_1/g_1, \ldots, f_m/g_m) \in \mathbb{K}[X_1, \ldots, X_n].$$

Next, we state the following:

Lemma 6.11 *Let V be an irreducible affine algebraic variety W be an affine algebraic variety and $\varphi \colon V - - \to W$ be a rational map. Suppose that O is an open subset of W. Then, $\{x \in \mathrm{dom}\varphi \mid \varphi(x) \in O\}$ is a possibly empty open subset of $\mathrm{dom}\varphi$.*

Proof Suppose that $a \in \mathrm{dom}\varphi$ and $\varphi(a) \in O$. Then by Lemma 6.3, we see that there exists $h \in \mathbb{K}[Y_1, \ldots, Y_m]$ such that $\varphi(a) \in W \setminus \mathbb{V}(h) \subset O$. Set $\varphi = (\varphi_1, \ldots, \varphi_m)$ and take representative (f_i, g_i) of φ_i with $g_i(a) \neq 0$ for $1 \leq i \leq m$. By Lemma 6.10, we can take a positive integer d such that

$$(g_1 \ldots g_m)^d h(f_1/g_1, \ldots, f_m/g_m) \in \mathbb{K}[X_1, \ldots, X_n].$$

Set $f = (g_1 \ldots g_m)^d h(f_1/g_1, \ldots, f_m/g_m)$. Then for any $x \in \mathrm{dom}\varphi$ with $(g_1 \ldots g_m)(x) \neq 0$,

$$h(\varphi(x)) \neq 0 \iff f(x) \neq 0.$$

Thus, $a \in \mathrm{dom}\varphi \setminus \mathbb{V}(g_1 \ldots g_m f) \subset \{x \in \mathrm{dom}\varphi \mid \varphi(x) \in O\}$. Since a is an arbitrary element of $\{x \in \mathrm{dom}\varphi \mid \varphi(x) \in O\}$, we see that $\{x \in \mathrm{dom}\varphi \mid \varphi(x) \in O\}$ is an open subset of $\mathrm{dom}\varphi$.

Lemma 6.12 *Let V be an irreducible affine algebraic variety in $\mathbb{K}^n$ and W be an irreducible affine algebraic variety in $\mathbb{K}^m$. Suppose that $\varphi = (\varphi_1, \ldots, \varphi_m)$ is a rational map from V to W and ψ is a rational function on W. If $\{x \in \mathrm{dom}\varphi \mid \varphi(x) \in \mathrm{dom}\psi\} \neq \emptyset$, then there exists a rational function χ on V such that $\mathrm{dom}\chi \supset \{x \in \mathrm{dom}\varphi \mid \varphi(x) \in \mathrm{dom}\psi\}$ and $\chi(a) = \psi(\varphi(a))$ for any $a \in \{x \in \mathrm{dom}\varphi \mid \varphi(x) \in \mathrm{dom}\psi\}$.*

Proof We use $\mathbb{K}[X_1, \ldots, X_n]$ to explain the facts concerning affine algebraic varieties in $\mathbb{K}^n$ and $\mathbb{K}[Y_1, \ldots, Y_m]$ in $\mathbb{K}^m$. Let x_1 be an arbitrary element of $\{x \in \mathrm{dom}\varphi \mid \varphi(x) \in \mathrm{dom}\psi\}$. Take a representative (h_1, h_2) of ψ such that $h_2(\varphi(x_1)) \neq 0$ and a representative (f_i, g_i) of φ_i such that $g_i(x_1) \neq 0$ for $1 \leq i \leq m$. Then there exists a positive integer d such that

$$(g_1 \ldots g_m)^d h_i(f_1/g_1, \ldots, f_m/g_m) \in \mathbb{K}[X_1, \ldots, X_m]$$

for $i = 1, 2$ by Lemma 6.10. Set $h_i' = (g_1 \ldots g_m)^d h_i(f_1/g_1, \ldots, f_m/g_m)$ for $i = 1$, 2. Then $h_2'(x_1) \neq 0$ and

$$\psi(\varphi(y)) = h_1'(y)/h_2'(y)$$

for any $y \in \mathrm{dom}\varphi$ with $\varphi(y) \in \mathrm{dom}\psi$ and $h_2'(y) \neq 0$. Let χ_1 be the rational function on V represented by (h_1', h_2'). Set $O_1 = \{x \in \mathrm{dom}\varphi \mid \varphi(x) \in \mathrm{dom}\psi\} \setminus \mathbb{V}(h_2')$. Then, O_1 is an open neighborhood of x_1 such that

$$\chi_1(y) = \psi(\varphi(y)) \quad \text{for any } y \in O_1.$$

Let x_2 be another element of $\{x \in \mathrm{dom}\varphi \mid \varphi(x) \in \mathrm{dom}\psi\}$. Then by the same argument, we see that there exists an open neighborhood O_2 of x_2 and a rational function χ_2 on V such that

$$\chi_2(y) = \psi(\varphi(y)) \quad \text{for any } y \in O_2.$$

Then,
$$\chi_1(y) = \chi_2(y) \quad \text{for any } y \in O_1 \cap O_2.$$

Since V is irreducible, $O_1 \cap O_2$ is a dense open subset of V. Thus, $\chi_1 = \chi_2$ by Lemma 6.9. Set $\chi = \chi_1$. Then, χ is a rational function on V and $\chi(a) = \psi(\varphi(a))$ for any $a \in \{x \in \text{dom}\varphi \mid \varphi(x) \in \text{dom}\psi\}$.

Definition 6.9 In the setting of Lemma 6.12, we denote χ as $\psi \circ \varphi$ and call it the composition of φ and ψ.

Definition 6.10 Let V_i be irreducible affine algebraic varieties in $\mathbb{K}^{n_i}$ for $i = 1, 2$, W be an affine algebraic variety in $\mathbb{K}^m$, $\varphi: V_1--\rightarrow V_2$ and $\psi: V_2--\rightarrow W$ be rational maps. Suppose that there exists $x \in \text{dom}\varphi$ such that $\varphi(x) \in \text{dom}\psi$. Then, we can naturally define the composition $\psi \circ \varphi$ of φ and ψ: set $\psi = (\psi_1, \ldots, \psi_m)$ and we define $\psi \circ \varphi = (\psi_1 \circ \varphi, \ldots, \psi_m \circ \varphi)$.

Remark 6.5 When considering the composition of rational maps φ and ψ, it is essential that $\text{Im}\varphi \cap \text{dom}\psi \neq \emptyset$.

6.4 Standard Form of "Quasi-Tall" Tensors

In ten Berge (2000), ten Berge called an $I \times J \times K$ tensor with $I \geq J \geq K \geq 2$ and $JK - J < I < JK$ a tall tensor. For convenience of notation, we rotate such tensors and provide the following definitions:

Definition 6.11 If $2 \leq m \leq n$ and $(m-1)n < p \leq mn$, we call an $n \times p \times m$-tensor a quasi-tall tensor.

Let m, n, and p be integers with $2 \leq m \leq n$ and $(m-1)n < p \leq mn$ and let $T = (T_1; \ldots; T_m)$ be an $n \times p \times m$ tensor. Thus, T is a quasi-tall tensor by Definition 6.11. Set $l := p - (m-1)n$ and $l' := n - l = mn - p$.

Definition 6.12 We say that T is of the standard form if
$$T_k = (O_{n \times (k-1)n}, E_n, O_{n \times (p-kn)})$$

for $1 \leq k \leq m - 1$ and

$$T_m = \left(\begin{pmatrix} M \\ O_{l \times (m-2)n} \end{pmatrix}, O_{n \times l}, E_n \right)$$

for some $l' \times (m-2)n$ matrix M.

Now we state the main result of this section.

Theorem 6.8 *Let m, n, and p be integers with $2 \leq m \leq n$ and $(m-1)n < p \leq mn$.
Then, there exist rational maps*

$$\varphi_1^{n \times p \times m} : \mathbb{K}^{n \times p \times m} \dashrightarrow GL(n, \mathbb{K})$$
$$\varphi_2^{n \times p \times m} : \mathbb{K}^{n \times p \times m} \dashrightarrow GL(p, \mathbb{K})$$

such that

$$\varphi_1^{n \times p \times m}(T) T \varphi_2^{n \times p \times m}(T)$$

*is of standard form for any $T \in \operatorname{dom}\varphi_1^{n \times p \times m} \cap \operatorname{dom}\varphi_2^{n \times p \times m}$ and for any $n \times p \times m$
tensor S of standard form*

$$S \in \operatorname{dom}\varphi_1^{n \times p \times m} \cap \operatorname{dom}\varphi_2^{n \times p \times m}, \quad \varphi_1^{n \times p \times m}(S) = E_n \quad and \quad \varphi_2^{n \times p \times m}(S) = E_p.$$

In order to prove this theorem, we prepare the following lemma which is easily
proved. We use the notations, $M_{\leq j}$ ($M^{\leq i}$, $_{j<}M$, $^{i<}M$, resp.) which denote the $m \times j$
($i \times n$, $m \times (n-j)$, $(m-i) \times n$, resp.) matrix consisting of the first (first, last, last,
resp.) j (i, $n-j$, $m-i$, resp.) columns (rows, columns, rows, resp.) of M for an
$m \times n$ matrix M. We also use the same notations for tensors.

Lemma 6.13 *Let*

$$F : \mathbb{K}^{n \times p \times m} \rightarrow GL(p, \mathbb{K})$$

be the regular map defined by

$$F(T) = \begin{pmatrix} E_{(m-2)n} & O \\ -\begin{pmatrix} T_{m-1} \\ {}_{l'<}T_m \end{pmatrix}_{\leq(m-2)n} & E_{n+l} \end{pmatrix},$$

where $T = (T_1; \ldots; T_m) \in \mathbb{K}^{n \times p \times m}$. Then, if

$$T_{m-1} = (M, E_n, O_{n \times l}) \quad and \quad T_m = (M', O_{n \times l}, E_n),$$

where M and M' are $n \times (m-2)n$ matrices, then

$$T_{m-1}F(T) = (O_{n \times (m-2)n}, E_n, O_{n \times l}) \quad and \quad T_m F(T) = \left(\begin{pmatrix} M'' \\ O_{l \times (m-2)n} \end{pmatrix}, O_{n \times l}, E_n \right),$$

where M'' is an $l' \times (m-2)n$ matrix. Moreover, if

$$T_{m-1} = (O_{n \times (m-2)n}, E_n, O_{n \times l}) \quad and \quad T_m = \left(\begin{pmatrix} M'' \\ O_{l \times (m-2)n} \end{pmatrix}, O_{n \times l}, E_n \right),$$

then $F(T) = E_p$.

Proof of Theorem 6.8 We define the rational map

$$\chi_0 \colon \mathbb{K}^{n \times p \times m} \text{---} \to GL(p, \mathbb{K})$$

by

$$\chi_0(T) = \begin{pmatrix} \mathrm{fl}_2(T)^{\leq (m-2)n} \\ O_{(n+l) \times (m-2)n} \ E_{n+l} \end{pmatrix}^{-1}.$$

Set

$$S^{(1)} = (S_1^{(1)}; \ldots; S_m^{(1)}) := T\chi_0(T)$$

for any $n \times p \times m$ tensor T with $T \in \mathrm{dom}\chi_0$. Then,

$$S_k^{(1)} = (O_{n \times (k-1)n}, E_n, O_{n \times (p-kn)})$$

for $1 \leq k \leq m - 2$. We also define rational maps

$$\chi_1 \colon \mathbb{K}^{n \times p \times m} \text{---} \to GL(n, \mathbb{K})$$
and
$$\chi_2 \colon \mathbb{K}^{n \times p \times m} \text{---} \to GL(p, \mathbb{K})$$

by

$$\chi_1(T) = \varphi_1^{n \times (n+l)} \left({}_{(m-2)n <} \left(S_{m-1}^{(1)}; S_m^{(1)} \right) \right)$$

and

$$\chi_2(T) = \begin{pmatrix} \chi_1(T)^{-1} & & & \\ & \ddots & & \\ & & \chi_1(T)^{-1} & \\ & & & \varphi_2^{n \times (n+l)} {}_{(m-2)n <} ((S_{m-1}^{(1)}; S_m^{(1)}))) \end{pmatrix},$$

where $\varphi_1^{n \times (n+l)}$ and $\varphi_2^{n \times (n+l)}$ are the ones in Theorem 6.4, and set

$$S^{(2)} = (S_1^{(2)}; \ldots; S_m^{(2)}) := \chi_1(T) S^{(1)} \chi_2(T).$$

Then

$$S_k^{(2)} = (O_{n \times (k-1)n}, E_n, O_{n \times (p-kn)})$$

for $1 \leq k \leq m - 2$,

$$S_{m-1}^{(2)} = (M, E_n, O_{n \times l}) \quad \text{and} \quad S_m^{(2)} = (M', O_{n \times l}, E_n),$$

where M and M' are $n \times (m-2)n$ matrices.

Set

$$\varphi_1^{n \times p \times m}(T) = \chi_1(T) \quad \text{and} \quad \varphi_2^{n \times p \times m}(T) = \chi_0(T)\chi_2(T)F(S^{(2)}),$$

where F is the regular map of Lemma 6.13. Then $\varphi_1^{n \times p \times m}$ and $\varphi_2^{n \times p \times m}$ satisfy the required conditions. ∎

6.5 Ranks of Quasi-Tall Tensors of Standard Form

In this section, we prove that an $n \times p \times m$ quasi-tall tensor of standard form has rank p. As a corollary, we give another proof of the result of ten Berge (2000), i.e., $p \times n \times m$ real tensors with $p \geq n \geq m \geq 2$ and $mn - n < p < mn$ have unique typical rank p.

Theorem 6.9 *Let $\mathbb{K}$ be an infinite field, m, n, and p be integers with $2 \leq m \leq n$ and $(m - 1)n < p \leq mn$ and S be an $n \times p \times m$ tensor of standard form (see Definition 6.12 for the definition of standard form). Then, $\text{rank} S = p$.*

In order to prove this theorem, we prepare some notations. Set $l := p - (m - 1)n$ and $l' := n - l = mn - p$. For an $l' \times (m - 2)n$ matrix $W = (W_1, \ldots, W_{m-2})$, where W_k is an $l' \times n$ matrix for $1 \leq k \leq m - 2$, and $c = (c_1, \ldots, c_m) \in \mathbb{K}^{1 \times nm}$, where $c_k \in \mathbb{K}^{1 \times n}$ for $1 \leq k \leq m$, we set

$$M(W, c) = M(x, T, W, c)$$
$$= x_1 \begin{pmatrix} W_1 \\ T_1 \\ c_1 \end{pmatrix} + \cdots + x_{m-2} \begin{pmatrix} W_{m-2} \\ T_{m-2} \\ c_{m-2} \end{pmatrix}$$
$$+ x_{m-1} \begin{pmatrix} O_{l' \times l} \ E_{l'} \\ T_{m-1} \\ c_{m-1} \end{pmatrix} + x_m \begin{pmatrix} -E_{l'} \ O_{l' \times l} \\ T_m \\ c_m \end{pmatrix},$$

where $x = (x_1, \ldots, x_m)$ is a row vector of indeterminates and $T = (T_1; \ldots; T_m) = (t_{ijk})$ is an $(l - 1) \times n \times m$ tensor of indeterminates. Note that when $p = (m-1)n+1$, i.e., when $l = 1$,

$$M(W, c) = x_1 \begin{pmatrix} W_1 \\ c_1 \end{pmatrix} + \cdots + x_{m-2} \begin{pmatrix} W_{m-2} \\ c_{m-2} \end{pmatrix}$$
$$+ x_{m-1} \begin{pmatrix} O_{l' \times l} \ E_{l'} \\ c_{m-1} \end{pmatrix} + x_m \begin{pmatrix} -E_{l'} \ O_{l' \times l} \\ c_m \end{pmatrix}.$$

We also set

$$g(x, T, W, c) = \det M(x, T, W, c),$$

which is a polynomial with variables x_1, ..., x_m, and $\{t_{ijk}\}$.

Lemma 6.14 *If* $(c_m)_{\leq l'} = 0$ *and* $g(x, T, W, c)$ *is a zero polynomial, then* $c = 0$.

Proof Set $c_k = (c_{k1}, \ldots, c_{kn})$ for $1 \leq k \leq m$. By seeing the coefficient of

$$x_m^n \prod_{i=l'+1}^{r-1} t_{i-l',i,m} \prod_{i=r}^{n-1} t_{i-l',i+1,m},$$

we see that $c_{rm} = 0$ for $l' + 1 \leq r \leq n$ (we define the empty product to be 1). Since $(c_m)_{\leq l'} = 0$ by assumption, we see that $c_m = 0$.

Thus by seeing the coefficient of

$$x_k x_m^{n-1} \prod_{i=l'+1}^{r-1} t_{i-l',i,m} \prod_{i=r}^{n-1} t_{i-l',i+1,m},$$

we see that $c_{rk} = 0$ for $1 \leq k \leq m - 1$ and $l' + 1 \leq r \leq n$.

Next by seeing the coefficient of

$$x_{m-1}^n \prod_{i=1}^{r-1} t_{i,i,m-1} \prod_{i=r}^{l-1} t_{i,i+1,m-1},$$

we see that $c_{r,m-1} = 0$ for $1 \leq r \leq l$. Further, when $l < l'$, by seeing the coefficient of

$$x_{m-1}^{n-r+1} x_m^{r-1} \prod_{i=1}^{l-1} t_{i,i+r-l,m},$$

we see that $c_{r,m-1}$ for $l + 1 \leq r \leq l'$. Thus, we see that $c_{m-1} = 0$.

Finally, by seeing the coefficient of

$$x_k x_{m-1}^{n-1} \prod_{i=1}^{r-1} t_{i,i,m-1} \prod_{i=r}^{l-1} t_{i,i+1,m-1},$$

we see that $c_{rk} = 0$ for $1 \leq k \leq m - 2$ and $1 \leq r \leq l$. Further, when $l < l'$, by seeing the coefficient of

$$x_k x_{m-1}^{n-r} x_m^{r-1} \prod_{i=1}^{l-1} t_{i,i+r-l,m},$$

we see that $c_{rk} = 0$ for $1 \leq k \leq m - 2$ and $l + 1 \leq r \leq l'$.

Definition 6.13 For an $l' \times (m - 2)n$ matrix W, $(l - 1) \times n \times m$ tensor U, and $u = (u_1, \ldots, u_m) \in \mathbb{K}^{1 \times m}$, we define $\psi(u, U, W) \in \mathbb{K}^n$ whose ith entry is the (n, i)-cofactor of $M(u, U, W, 0)$. We also define

$$\hat{\psi}(\boldsymbol{u}, U, W) := \begin{pmatrix} u_1\psi(\boldsymbol{u}, U, W) \\ \vdots \\ u_{m-1}\psi(\boldsymbol{u}, U, W) \\ u_m{}^{l'<}\psi(\boldsymbol{u}, U, W) \end{pmatrix} \in \mathbb{K}^p.$$

Lemma 6.15 *Let W be an $l' \times (m-2)n$ matrix. Then, the $\mathbb{K}$ vector subspace of $\mathbb{K}^p$ generated by $\{\hat{\psi}(\boldsymbol{u}, U, W) \mid \boldsymbol{u} \in \mathbb{K}^{1\times m}, U \in \mathbb{K}^{(l-1)\times n\times m}\}$ is $\mathbb{K}^p$.*

Proof Suppose that $\boldsymbol{d} \in \mathbb{K}^{1\times p}$ and $\boldsymbol{d}\hat{\psi}(\boldsymbol{u}, U, W) = 0$ for any $\boldsymbol{u} \in \mathbb{K}^{1\times m}$ and $U \in \mathbb{K}^{(l-1)\times n\times m}$. Set $\boldsymbol{d} = (d_1, \ldots, d_p)$ and define

$$\boldsymbol{c} = (d_1, \ldots, d_{(m-1)n}, 0, \ldots, 0, d_{(m-1)n+1}, \ldots, d_p) \in \mathbb{K}^{1\times mn}$$

by inserting l' zeros. Then by the definition of $\hat{\psi}(\boldsymbol{u}, U, W)$, we see that

$$\boldsymbol{d}\hat{\psi}(\boldsymbol{u}, U, W) = \det M(\boldsymbol{u}, U, W, \boldsymbol{c}) = g(\boldsymbol{u}, U, W, \boldsymbol{c}).$$

Since $\mathbb{K}$ is an infinite field and $\boldsymbol{d}\hat{\psi}(\boldsymbol{u}, U, W) = 0$ for any $\boldsymbol{u} \in \mathbb{K}^{1\times m}$ and $U \in \mathbb{K}^{(l-1)\times n\times n}$, we see that $g(\boldsymbol{x}, T, W, \boldsymbol{c})$ is a zero polynomial, where $\boldsymbol{x} = (x_1, \ldots, x_m)$ is a vector of indeterminates and T is an $(l-1) \times n \times m$ tensor of indeterminates. Thus by Lemma 6.14, we see that $\boldsymbol{d} = \boldsymbol{0}$. Therefore, the $\mathbb{K}$-vector space generated by $\{\hat{\psi}(\boldsymbol{u}, U, W) \mid \boldsymbol{u} \in \mathbb{K}^{1\times m}, U \in \mathbb{K}^{(l-1)\times n\times m}\}$ is $\mathbb{K}^p$.

Proof of Theorem 6.9

Let $S = (S_1; \ldots; S_m)$ be an $n \times p \times m$ tensor of standard form. Since $\mathrm{fl}_2(S)^{\leq p} = E_p$, we see that $\mathrm{rank}\, S \geq p$.

In order to prove the opposite inequality, we set

$$S_m = \begin{pmatrix} W \\ O_{l\times(m-2)n} \end{pmatrix}, O_{n\times l}, E_n \end{pmatrix},$$

where W is an $l'\times(m-2)n$ matrix. Since $\{\hat{\psi}(\boldsymbol{u}, U, W) \mid \boldsymbol{u} \in \mathbb{K}^{1\times m}, U \in \mathbb{K}^{(l-1)\times n\times m}\}$ generates $\mathbb{K}^p$ by Lemma 6.15, we can take $\boldsymbol{u}_1, \ldots, \boldsymbol{u}_p \in \mathbb{K}^{1\times m}$ and $U_1, \ldots, U_p \in \mathbb{K}^{(l-1)\times n\times m}$ such that

$$\left(\hat{\psi}(\boldsymbol{u}_1, U_1, W), \ldots, \hat{\psi}(\boldsymbol{u}_p, U_p, W)\right)$$

is a nonsingular $p \times p$ matrix.

We denote this matrix as N, and set $Q = N^{-1}$, $\boldsymbol{u}_j = (u_{j1}, u_{j2}, \ldots, u_{jm})$ for $1 \leq j \leq p$, $D_k = \mathrm{Diag}(u_{1k}, u_{2k}, \ldots, u_{pk})$ for $1 \leq k \leq m$, $\boldsymbol{a}_j = \psi(\boldsymbol{u}_j, U_j, W)$ for $1 \leq j \leq p$ and $A = (\boldsymbol{a}_1, \ldots, \boldsymbol{a}_p)$. Then, since

$$
\begin{pmatrix} u_{j1}a_j \\ \vdots \\ u_{j,m-1}a_j \\ u_{j,m}{}^{l'<}a_j \end{pmatrix} = \hat{\psi}(\boldsymbol{u}_j, U_j, W)
$$

for $1 \le j \le p$, we see that

$$
\begin{pmatrix} AD_1 \\ \vdots \\ AD_{m-1} \\ {}^{l'<}AD_m \end{pmatrix} = N.
$$

Therefore,

$$
\begin{pmatrix} AD_1Q \\ \vdots \\ AD_{m-1}Q \\ {}^{l'<}AD_mQ \end{pmatrix} = NQ = E_p.
$$

In other words,

$$
AD_kQ = S_k \quad \text{for } 1 \le k \le m-1 \tag{6.5.1}
$$

and

$$
{}^{l'<}AD_mQ = {}^{l'<}S_m. \tag{6.5.2}
$$

Now consider $A^{\le l'}D_mQ$. Since the ith entry of $\boldsymbol{a}_j = \psi(\boldsymbol{u}_j, U_j, W)$ is the (n, i)-cofactor of $M(\boldsymbol{u}_j, U_j, W, \mathbf{0})$, we see that

$$
M(\boldsymbol{u}_j, U_j, W, \mathbf{0})\boldsymbol{a}_j = \mathbf{0} \quad \text{for } 1 \le j \le p. \tag{6.5.3}
$$

Set $U_j = (U_{j1}; \ldots; U_{jm})$ for $1 \le j \le p$. Then, since

$$
M(\boldsymbol{u}_j, U_j, M, \mathbf{0}) = u_{j1} \begin{pmatrix} W_1 \\ U_{j1} \\ \mathbf{0} \end{pmatrix} + \cdots + u_{j,m-2} \begin{pmatrix} W_{m-2} \\ U_{j,m-2} \\ \mathbf{0} \end{pmatrix}
$$
$$
+ u_{j,m-1} \begin{pmatrix} O_{l'\times l} \ E'_l \\ U_{j,m-1} \\ \mathbf{0} \end{pmatrix} + u_{j,m} \begin{pmatrix} -E_{l'} \ O_{l'\times l} \\ U_{j,m} \\ \mathbf{0} \end{pmatrix},
$$

by seeing the first l' rows of (6.5.3), we see that

$$
(u_{j1}W_1 + \cdots + u_{j,m-2}W_{m-2} + u_{j,m-1}(O_{l'\times l} \ E_{l'}))\boldsymbol{a}_j = u_{jm}\boldsymbol{a}_j^{\le l'}
$$

for $1 \leq j \leq p$. Therefore,

$$W_1 A D_1 + \cdots + W_{m-2} A D_{m-2} + (O_{l' \times l} \; E_{l'}) A D_{m-1} = (A D_m)^{\leq l'}.$$

Since

$$S_m^{\leq l'} = (W_1, \ldots, W_{m-2}, O_{l' \times l}, E_{l'}, O_{l' \times l}),$$

we see that

$$S_m^{\leq l'} N = A^{\leq l'} D_m.$$

Therefore,

$$A^{\leq l'} D_m Q = S_m^{\leq l'}.$$

By this fact and Eq. (6.5.2), we see that

$$A D_m Q = S_m.$$

Thus, by (6.5.1), we see that $\operatorname{rank} S \leq p$. ∎

By Theorems 6.8 and 6.9, we see the following fact:

Corollary 6.3 *Let $\mathbb{K}$ be an infinite field, m, n, and p be integers with $2 \leq m \leq n$ and $(m-1)n < p \leq mn$. Then there exists a dense open subset $\mathcal{O}$ of $\mathbb{K}^{n \times p \times m}$ such that for any $T \in \mathcal{O}$, $\operatorname{rank} T = p$.*

Proof Set $\mathcal{O} = \operatorname{dom}\varphi_1^{n \times p \times m} \cap \operatorname{dom}\varphi_2^{n \times p \times m}$, where $\varphi_1^{n \times p \times m}$ and $\varphi_2^{n \times p \times m}$ are rational maps of Theorem 6.8. If $T \in \mathcal{O}$, then $\operatorname{rank} T = \operatorname{rank} \varphi_1^{n \times p \times m}(T) T \varphi_2^{n \times p \times m}(T)$. Since $\varphi_1^{n \times p \times m}(T) T \varphi_2^{n \times p \times m}(T)$ is an $n \times p \times m$ tensor of standard form, $\operatorname{rank} \varphi_1^{n \times p \times m}(T) T \varphi_2^{n \times p \times m}(T) = p$ by Theorem 6.9. ∎

In particular, we see the following fact:

Corollary 6.4 (ten Berge 2000, Result 2) *Suppose that m, n, and p are integers with the condition of the above corollary. Then, p is the unique typical rank of $n \times p \times m$ tensors over the real number field.*

6.6 Other Cases

Here, in the final section of this chapter, we consider typical ranks of 3-tensors that are not quasi-tall. First, we state the following result:

Proposition 6.1 *If $p > mn$, then $\operatorname{trank}_{\mathbb{R}}(n, p, m) = \{mn\}$.*

Proof Let T be an $n \times p \times m$ tensor. Then, $\operatorname{fl}_2(T)$ is an $nm \times p$ matrix. Since $nm < p$, the rank of $\operatorname{fl}_2(T)$ is almost always nm. Therefore, $\operatorname{rank} T$ is almost always greater than or equal to nm.

On the other hand, let M_{ij} be an element of $\mathbb{R}^{n \times 1 \times m}$ such that $\mathrm{fl}_1(M_{ij}) = E_{ij}$, the matrix unit. Then, since every column of T is a linear combination of M_{ij} and rank $M_{ij} = 1$, we see that rank of T is less than or equal to the number of M_{ij}'s, i.e., nm. Thus, rank $T \leq nm$.

Next, we cite the result of (Catalisano et al. 2008), Theorem 2.4 and Remark 2.5, which, in our language, includes the following fact:

Theorem 6.10 *Let m, n, and p be positive integers with* $3 \leq m \leq n$ *and* $(m - 1)(n - 1) < p \leq mn$. *Then,* $\mathrm{grank}_\mathbb{C}(n, p, m) = p$.

Thus, it follows that min $\mathrm{trank}_\mathbb{R}(n, p, m) = p$ by Theorem 6.6, if m, n, and p satisfy the conditions of Theorem 6.10.

Here we cite our following recent result:

Theorem 6.11 (Sumi et al. 2013, 2015a, b) *Let m, n, and p be integers with* $3 \leq m \leq n$ *and* $(m - 1)(n - 1) + 2 \leq p \leq (m - 1)n$. *Then*

$$\mathrm{trank}_\mathbb{R}(n, p, m) = \begin{cases} \{p, p + 1\} & \text{if } m\#n \leq mn - p, \\ \{p\} & \text{if } m\#n > mn - p, \end{cases}$$

where m#n is the minimum integer l such that there exists a nonsingular bilinear map $\mathbb{R}^m \times \mathbb{R}^n \to \mathbb{R}^l$ *(see Chap. 4).*

Finally, we cite the following fact:

Theorem 6.12 (Strassen 1983) *Let* $A \in \mathbb{K}^{n \times n \times 3}$. *If n is odd, then* $\mathrm{grank}(n, n, 3) = (3n + 1)/2$; *otherwise* $\mathrm{grank}(n, n, 3) = 3n/2$.

Chapter 7
Global Theory of Tensor Ranks

The generic rank is considered under the complex number field, and it corresponds with the dimension of the secant variety. In this chapter, we introduce known results and discuss the typical rank via the Jacobi criterion.

7.1 Overview

Let $\mathbb{K}$ be an algebraically closed field or $\mathbb{R}$. Let $f = (f_1, \ldots, f_k)$. We identify $T_{\mathbb{K}}(f)$ with $\mathbb{K}^{f_1 f_2 \cdots f_k}$. We consider the tensor product map $\Phi_1 \colon \mathbb{K}^{f_1} \times \mathbb{K}^{f_2} \times \cdots \times \mathbb{K}^{f_k} \to T_{\mathbb{K}}(f)$ defined as

$$(u_1, u_2, \ldots, u_k) \mapsto u_1 \otimes u_2 \otimes \cdots \otimes u_k.$$

It induces a map from $\mathbb{P}^{f_1-1} \times \mathbb{P}^{f_2-1} \times \cdots \times \mathbb{P}^{f_k-1}$ to $\mathbb{P}^{f_1 \cdots f_k-1}$, where $\mathbb{P}^n$ is the n-dimensional projective space over $\mathbb{K}$, which is a quotient space of $\mathbb{K}^{n+1} \setminus \{0\}$ by the equivalence relation $\sim$ defined by $(x_1, \ldots, x_{n+1}) \sim (y_1, \ldots, y_{n+1})$ if $(x_1, \ldots, x_{n+1}) = (cy_1, \ldots, cy_{n+1})$ for some nonzero element $c \in \mathbb{K}$. The image of Φ_1 is an algebraic variety. We call it the *Segre variety* of the format f (cf. Bürgisser et al. 1997, Chap. 20, Harris 1992, Lecture 9). The image of the summation map

$$\Phi_r^{\mathbb{K}, f} \colon (\mathbb{K}^{f_1} \times \mathbb{K}^{f_2} \times \cdots \times \mathbb{K}^{f_k})^r \to T_{\mathbb{K}}(f), \ (u_1, u_2, \ldots, u_r) \mapsto \sum_{j=1}^{r} \Phi_1(u_j),$$

denoted by $S_r(f, \mathbb{K})$ or $S_r(f)$, consists of the tensors in $\mathbb{K}^f$ of rank $\leq r$. We simply write $\Phi_r^{\mathbb{K}, f}$ as Φ_r. The Zariski closure, denoted by $\Sigma_r(f, \mathbb{K})$ or $\Sigma_r(f)$, of $S_r(f, \mathbb{K})$ is called the *secant variety* of the format f. The secant variety $\Sigma_r(f, \mathbb{C})$ is irreducible and consists of all tensors of $\mathbb{C}^f$ with border rank at most r (cf. Lickteig 1985). The maximal rank max.rank$_{\mathbb{K}}(f)$ of $T_{\mathbb{K}}(f)$ is characterized as follows.

Remark 7.1 max.rank$_{\mathbb{K}}(f) = \min\{r \mid S_r(f) = T_{\mathbb{K}}(f)\}$.

© The Author(s) 2016
T. Sakata et al., *Algebraic and Computational Aspects of Real Tensor Ranks*,
JSS Research Series in Statistics, DOI 10.1007/978-4-431-55459-2_7

An integer r is called a *typical rank* of the set $T_{\mathbb{R}}(f)$ if $S_r(f, \mathbb{R}) \setminus S_{r-1}(f, \mathbb{R})$ includes a nonempty Euclidean open set. Let $\text{trank}_{\mathbb{R}}(f)$ denote the set of typical ranks of $T_{\mathbb{R}}(f)$.

A *semi-algebraic set* in $\mathbb{R}^n$ is a finite union of sets defined by a finite number of polynomial equations of the form $p(x_1, \ldots, x_n) = 0$ and inequalities of the form $q(x_1, \ldots, x_n) > 0$. The set $T_{\mathbb{R}}(f)$ is a semi-algebraic set. For semi-algebraic sets A and B, the product $A \times B$ is also semi-algebraic. A finite union, finite intersection, complements, interiors, and closures of semi-algebraic sets are also semi-algebraic sets. According to the Tarski–Seidenberg principle, the set of semi-algebraic sets is closed under projection. Let $h: A \to B$ be a map between semi-algebraic sets. A map h is called *semi-algebraic* if its graph $\{(x, h(x)) \mid x \in A\}$ is a semi-algebraic subset of $A \times B$. If h is a polynomial map, then h is semi-algebraic. For a semi-algebraic map h, the image $h(S)$ of a semi-algebraic subset S of A is a semi-algebraic subset of B and the preimage $h^{-1}(T)$ of a semi-algebraic subset T of B is also a semi-algebraic subset of A. In particular, $S_r(f, \mathbb{R})$ and the set $S_r(f, \mathbb{R}) \setminus S_{r-1}(f, \mathbb{R})$ of all tensors of $T_{\mathbb{R}}(f)$ with rank r are semi-algebraic sets (see Bochnak et al. 1998, Chap. 2, Sect. 2 in detail).

Since Φ_r is a C^∞ map, we can consider the Jacobi $f_1 f_2 \cdots f_k \times r(f_1 + f_2 + \cdots + f_k)$ matrix, denoted by J_{Φ_r}, of the map Φ_r. We see that

$$J_{\Phi_1}(u_1, \ldots, u_k) = (E_{f_1} \otimes u_2 \otimes \cdots \otimes u_k, u_1 \otimes E_{f_2} \otimes \cdots \otimes u_k, \ldots, u_1 \otimes \cdots \otimes u_{k-1} \otimes E_{f_k})$$

for $(u_1, \ldots, u_k) \in \mathbb{K}^{f_1} \times \cdots \times \mathbb{K}^{f_k}$ and that

$$J_{\Phi_r}(x_1, \ldots, x_r) = (J_{\Phi_1}(x_1), J_{\Phi_1}(x_2), \ldots, J_{\Phi_1}(x_r))$$

where $x_1, \ldots, x_r \in \mathbb{K}^{f_1} \times \cdots \times \mathbb{K}^{f_k}$. We put

$$d_r(f) = \max_x \text{rank} J_{\Phi_r}(x).$$

Over the complex number field, the generic rank $\text{grank}(f)$ denotes the minimal integer r such that $d_r(f) = f_1 f_2 \cdots f_k$. Note that $\Sigma_{\text{grank}(f)}(f, \mathbb{C}) = \mathbb{C}^f$. Over the real number field $\mathbb{R}$, if $d_r(f) = f_1 f_2 \cdots f_k$, then r is greater than or equal to the generic rank $\text{grank}(f)$ which is equal to the minimal typical rank $\min.\text{trank}_{\mathbb{R}}(f)$, and the maximal typical rank $\max.\text{trank}_{\mathbb{R}}(f)$ is equal to the minimal integer r such that the Euclidean closure of $S_r(f, \mathbb{R})$ is equal to $T_{\mathbb{R}}(f)$.

Strassen (1983) and Lickteig (1985) introduced the idea of computing upper bounds on the typical rank via the Jacobi criterion and the splitting technique.

An integer r is called *small* if $\dim \Sigma_r(m, n, q) = r(m + n + q - 2)$, and *large* if $\Sigma_r(m, n, q) = mnq$. A format (m, n, q) is called *good* if $\dim \Sigma_r(m, n, q) = \min\{r(m+n+q-2), mnq\}$ for any r, and *perfect* if, in addition, $mnq/(m+n+q-2)$ is an integer. Let us call a format (m, n, q) *balanced* if $m - 1 \leq (n - 1)(q - 1)$, $n - 1 \leq (m - 1)(q - 1)$, and $q - 1 \leq (m - 1)(n - 1)$.

Let $2 \leq m \leq n \leq q \leq mn$ and $q_0 = (m-1)(n-1)+1$. In ten Berge 2000, a tensor with format (m, n, q) is called "tall" if $(m-1)n < q < mn$. A tensor with format (m, n, q) with $(m-1)n < q \leq mn$ has rank q with probability 1 (see Corollary 6.3). By considering the flattening $\mathbb{R}^{m \times n \times q} \to \mathbb{R}^{mn \times q}$, we see that $\min.\mathrm{trank}_{\mathbb{R}}(m, n, q) \geq \min.\mathrm{trank}_{\mathbb{R}}(mn, q) = q$. By the argument of the rank of the Jacobi matrix (see Theorem 7.8), $\min.\mathrm{trank}_{\mathbb{R}}(m, n, q) \leq q$ for $q_0 \leq q \leq mn$. Therefore, if $q_0 \leq q \leq mn$, then $\min.\mathrm{trank}_{\mathbb{R}}(m, n, q) = q$, and a tensor with format (m, n, q) has rank q with positive probability. In particular, (m, n, q_0) is perfect. Conversely, if $\min.\mathrm{trank}_{\mathbb{R}}(m, n, q) \leq q$, then $q \geq q_0$ since $mnq/(m+n+q-2) \leq \min.\mathrm{trank}_{\mathbb{R}}(m, n, q)$. If $q_0 + 2 \leq q \leq mn$, then $(q-1)(m+n+q-2) \geq mnq$ and (m, n, q) is not good, since $\dim \Sigma_{q-1}(m, n, q) < mnq$. (Bürgisser et al. 1997, Exercise 20.6).

Theorem 7.1 (Strassen 1983, Proposition 3.9 and Corollaries 3.10 and 3.11) *Suppose that (m, n, q) is a balanced format. Then, (m, n, q) is perfect provided that any of the following conditions is satisfied:*

$$q \text{ even, } 2n < m+n+q-2, \text{ and } 2mn/(m+n+q-2) \in \mathbb{Z}.$$
$$q/3 \in \mathbb{Z}, 3n \leq m+n+q-2, \text{ and } 3mn/(m+n+q-2) \in \mathbb{Z}.$$
$$3n \leq m+n+q-2 \text{ and } mn/(m+n+q-2) \in \mathbb{Z}.$$

In particular, if $n \not\equiv 2$ modulo 3, then $(n, n, n+2)$ is perfect; if $n \equiv 0$ modulo 3, then $(n-1, n, n)$ is perfect; and if $j \equiv 0$ modulo $2(\alpha+\beta+\gamma)$ and $1 \leq \alpha \leq \beta \leq \gamma$, then $(\alpha j, \beta j, \gamma j + 2)$ is perfect.

Let $A_1 = E_n$, $A_2 = \mathrm{Diag}(c_1, c_2, \ldots, c_n)$, and $A_3 = \begin{pmatrix} 0 & 1 & & & \\ & 0 & 1 & & \\ & & \ddots & \ddots & \\ & & & 0 & 1 \\ 1 & & & & 0 \end{pmatrix}$. If

$c_1, \ldots, c_n$ are distinct from each other, then

$$A_2 A_3 - A_3 A_2 = \begin{pmatrix} 0 & c_1 - c_2 & & & \\ & 0 & c_2 - c_3 & & \\ & & \ddots & \ddots & \\ & & & 0 & c_{n-1} - c_n \\ c_n - c_1 & & & & 0 \end{pmatrix}$$

has rank n and $\mathrm{brank}_{\mathbb{K}}(A_1; A_2; A_3) \geq \lceil \frac{3n}{2} \rceil$ by Theorem 5.15. Therefore,

$$\mathrm{grank}(n, n, 3) \geq \lceil \frac{3n}{2} \rceil.$$

Although there are infinite many good formats, $(n, n, 3)$ is not good if n is odd.

Theorem 7.2 (Strassen 1983, Theorem 4.6 and Proposition 4.7) *If n is odd, then* grank$(n, n, 3) = (3n + 1)/2$; *otherwise,* grank$(n, n, 3) = 3n/2$.

Theorem 7.3 (Lickteig 1985, Corollary 4.5) grank$(n, n, n) = \lceil \frac{n^3}{3n-2} \rceil$ *if* $n \neq 3$. dim $\Sigma_r(n, n, n) = \min\{r(3n - 2), n^3\}$.

Thus, $(n, n, 3)$ is good if and only if n is even, and (n, n, n) is good if $n \neq 3$.

Theorem 7.4 (Strassen 1983, Proposition 4.7) *Let* $2 \leq m \leq n \leq m + q - 2$, $q \leq (m - 1)(n - 1) + 1$, *and q be even.*

$$\frac{mnq}{m + n + q - 2} \leq \text{grank}(m, n, q) < \frac{mnq}{m + n + q - 2} + \frac{q}{2}$$

7.2 Jacobian Method

Let $\mathbb{K}$ be an algebraically closed field or $\mathbb{R}$. Let $f = (f_1, f_2, \ldots, f_k)$ be a format with $f_1, f_2, \ldots, f_k \geq 2$. The Jacobi matrix $J_{\Phi_r}(x_1, \ldots, x_r)$ forms

$$(J_{\Phi_1}(x_1), \ldots, J_{\Phi_1}(x_r)).$$

Since

$$\Phi_1(x) = \sum_{i=1}^{f_s} x_{s,i_s} \frac{\partial \Phi_1(f)(x)}{\partial x_{s,i_s}},$$

where $x = (x_{s,i_s})_{1 \leq s \leq k, 1 \leq i_s \leq f_s}$, we see that rank$(J_{\Phi_1}(x)) \leq \sum_{i=1}^{k} f_i - k + 1$, and then,

$$\text{rank}(J_{\Phi_r}(x_1, \ldots, x_r)) \leq r \left(\sum_{i=1}^{k} f_i - k + 1 \right).$$

Note that

$$d_1(f) < d_2(f) < \cdots < d_{\text{grank}(f)}(f) = f_1 \cdots f_k.$$

We put

$$Q(f) := \left\lceil \frac{f_1 f_2 \cdots f_k}{f_1 + f_2 + \cdots + f_k - k + 1} \right\rceil.$$

Proposition 7.1 (cf. Howell 1978, Theorem 12; Bürgisser et al. 1997, (20.4) Proposition (5)) $Q(f) \leq \text{grank}(f) \leq \text{max.rank}_{\mathbb{F}}(f)$.

Proof Recall that grank(f) is the minimal integer r such that $d_r(f) = f_1 f_2 \cdots f_k$. Since $\Sigma_{\text{max.rank}_{\mathbb{F}}(f)}(f) = \mathbb{F}^f$, we see that grank$(f) \leq \text{max.rank}_{\mathbb{F}}(f)$. If $r < Q(f)$, then $d_r(f) < f_1 f_2 \cdots f_k$. Thus, $Q(f) \leq \text{grank}(f)$.

Theorem 7.5 (Howell 1978, Theorem 10) *If R is a finite commutative ring with identity, then* $\max.\mathrm{rank}_R(m, n, p) \geq \lceil mnp/(m + n + p - 2)\rceil$. *If $\mathbb{K}$ is a finite field with q elements, then* $\max.\mathrm{rank}_R(m, n, p) \geq \lceil mnp/(m + n + p - 2\log_q(q - 1))\rceil$.

The typical rank of $T_{\mathbb{R}}(f)$ is not unique in general, e.g., 2 and 3 are typical ranks of $T_{\mathbb{R}}(2, 2, 2)$. The generic rank $\mathrm{grank}(f)$ is the minimal typical rank of $T_{\mathbb{R}}(f)$ (cf. Northcott 1980, Theorem 6.6). $S_r(f, \mathbb{R})$ is a semi-algebraic set (cf. de Silva and Lim 2008; Friedland 2012).

For a permutation $\tau \in \mathfrak{S}_k$, we see that

$$\max.\mathrm{rank}_{\mathbb{K}}(f_1, f_2, \ldots, f_k) = \max.\mathrm{rank}_{\mathbb{K}}(f_{\tau(1)}, f_{\tau(2)}, \ldots, f_{\tau(k)}),$$
$$\mathrm{grank}(f_1, f_2, \ldots, f_k) = \mathrm{grank}(f_{\tau(1)}, f_{\tau(2)}, \ldots, f_{\tau(k)}),$$
$$\mathrm{trank}_{\mathbb{R}}(f_1, f_2, \ldots, f_k) = \mathrm{trank}_{\mathbb{R}}(f_{\tau(1)}, f_{\tau(2)}, \ldots, f_{\tau(k)}).$$

We know the following upper bound for the maximal rank.

Proposition 7.2 (Proposition 1.2) $\max.\mathrm{rank}_{\mathbb{K}}(f_1, \ldots, f_k) \leq \min\{\frac{f_1 \cdots f_k}{f_j} \mid 1 \leq j \leq k\}$.

The following proposition is elementary.

Proposition 7.3 *The minimal typical rank r is characterized as $d_r(f) = f_1 \cdots f_k > d_{r-1}(f)$.*

Proof It is obvious by the definition.

Let $A \in T_{\mathbb{K}}(a_1, a_2, \ldots, a_s)$. For $B = (B_1; \ldots; B_{b_t}) \in T_{\mathbb{K}}(b_1, b_2, \ldots, b_t)$, if $s \geq t$ and $(B_1; \ldots; B_{b_s})$ has A as a sub-tensor, then $\mathrm{rank}_{\mathbb{K}}(B) \geq \mathrm{rank}_{\mathbb{K}}(B_1; \ldots; B_{b_s}) \geq \mathrm{rank}_{\mathbb{K}}(A)$.

Let $\max.\mathrm{trank}_{\mathbb{R}}(f)$ be the maximal typical rank of $T_{\mathbb{R}}(f)$. Recall that $\mathrm{grank}(f)$ is the minimal typical rank of $T_{\mathbb{R}}(f)$ (see Theorem 6.6). Let $\mathscr{T}_r(f, \mathbb{K})$ denote the set of all tensors of $\mathbb{K}^f$ with rank r. We have

$$S_r(f, \mathbb{K}) = \bigcup_{i=0}^{r} \mathscr{T}_i(f, \mathbb{K}).$$

Lemma 7.1 *$S_r(f, \mathbb{R})$ is a Euclidean dense subset of $T_{\mathbb{R}}(f)$ if and only if r is greater than or equal to the maximal typical rank of $T_{\mathbb{R}}(f)$.*

Let $a = (a_1, a_2, \ldots, a_s)$ and $b = (b_1, b_2, \ldots, b_t)$ with $s \geq t$ and $a_i \geq b_i$ for any $i \leq t$, let r be a nonnegative integer, and let $\pi: \mathbb{K}^a \to \mathbb{K}^b$ be a canonical projection. By definition, $\pi(S_r(a, \mathbb{K})) \subset S_r(b, \mathbb{K})$. The inclusion $\mathbb{K}^b \to \mathbb{K}^a$ by adding zero at the other elements implies that $S_r(b, \mathbb{K}) \subset \pi(S_r(a, \mathbb{K}))$. Therefore, $S_r(b, \mathbb{K}) = \pi(S_r(a, \mathbb{K}))$.

Proposition 7.4 *Let $a = (a_1, a_2, \ldots, a_s)$ and $b = (b_1, b_2, \ldots, b_t)$. If $s \geq t$ and $a_i \geq b_i$ for any $i \leq t$, then*

(1) $\mathrm{grank}(a) \geq \mathrm{grank}(b)$ *and*
(2) $\mathrm{max.trank}_{\mathbb{R}}(a) \geq \mathrm{max.trank}_{\mathbb{R}}(b)$.

Proof (1) Note that $S_{\mathrm{grank}(a)}(b, \mathbb{C}) = S_{\mathrm{grank}(a)}(a, \mathbb{C}) \cap \mathbb{C}^b$. Since it includes a Euclidean open set, $\mathrm{grank}(a) \geq \mathrm{grank}(b)$.

(2) $S_{\mathrm{max.trank}_{\mathbb{R}}(a)}(a, \mathbb{R})$ is a Euclidean dense subset of $T_{\mathbb{R}}(a)$. Let $\pi: T_{\mathbb{R}}(a) \rightarrow T_{\mathbb{R}}(b)$ be a canonical projection. Then, $\pi(S_{\mathrm{max.trank}_{\mathbb{R}}(a)}(a, \mathbb{R})) = S_{\mathrm{max.trank}_{\mathbb{R}}(a)}(b, \mathbb{R})$ is also a Euclidean dense subset of $T_{\mathbb{R}}(b)$. Therefore, $\mathrm{max.trank}_{\mathbb{R}}(a)$ is greater than or equal to $\mathrm{max.trank}_{\mathbb{R}}(b)$.

Proposition 7.5 *Let $a = (a_1, a_2, \ldots, a_s)$ and $b = (b_1, b_2, \ldots, b_t)$ with $s \geq t$ and $a_i \geq b_i$ for any $i \leq t$. $\mathrm{grank}(a) \geq \mathrm{max.trank}_{\mathbb{R}}(b)$ if and only if $\pi(S_{\mathrm{grank}(a)}(a, \mathbb{R}))$ is a Euclidean dense subset of $T_{\mathbb{R}}(b)$, where $\pi: T_{\mathbb{R}}(a) \rightarrow T_{\mathbb{R}}(b)$ is a canonical projection.*

Proof Recall that $\pi(\mathscr{T}_{\mathrm{grank}(a)}(a, \mathbb{R}))$ is a subset of $\mathscr{T}_{\mathrm{grank}(a)}(b, \mathbb{R})$. Thus, if it is a Euclidean dense subset of $T_{\mathbb{R}}(b)$, then

$$\mathrm{grank}(a) \geq \mathrm{max.trank}_{\mathbb{R}}(b)$$

by Lemma 7.1. Conversely, suppose that $\mathrm{grank}(a) \geq \mathrm{max.trank}_{\mathbb{R}}(b)$. Then,

$$T_{\mathbb{R}}(b) = \mathrm{cl}\, S_{\mathrm{max.trank}_{\mathbb{R}}(b)}(b, \mathbb{R}) \subset \mathrm{cl}\, S_{\mathrm{grank}(a)}(b, \mathbb{R}) = \mathrm{cl}\, \pi(S_{\mathrm{grank}(a)}(a, \mathbb{R})).$$

Lemma 7.2 *If $\mathrm{cl}\, S_{r+1}(f, \mathbb{R}) = \mathrm{cl}\, S_r(f, \mathbb{R})$, then $\mathrm{cl}\, S_{r+2}(f, \mathbb{R}) = \mathrm{cl}\, S_{r+1}(f, \mathbb{R})$.*

Proof An implication $S_{r+1}(f) \subset S_{r+2}(f)$ easily implies that $\mathrm{cl}\, S_{r+1}(f) \subset \mathrm{cl}\, S_{r+2}(f)$. Since

$$\begin{aligned} S_{r+2} &= S_{r+1}(f) + S_1(f) \subset \mathrm{cl}\, S_{r+1}(f) + S_1(f) \\ &= \mathrm{cl}\, S_r(f) + S_1(f) \subset \mathrm{cl}(S_r(f) + S_1(f)) = \mathrm{cl}\, S_{r+1}(f), \end{aligned}$$

we have $\mathrm{cl}\, S_{r+2}(f) \subset \mathrm{cl}\, S_{r+1}(f)$. Therefore, $\mathrm{cl}\, S_{r+1}(f) = \mathrm{cl}\, S_{r+2}(f)$.

Proposition 7.6 *$T_{\mathbb{R}}(f) \setminus S_r(f, \mathbb{R})$ and $S_{r+1}(f, \mathbb{R})$ include a nonempty Euclidean open set if and only if $r + 1 \in \mathrm{trank}_{\mathbb{R}}(f)$.*

Proof The if part follows from the fact that $\mathscr{T}_{r+1}(f, \mathbb{R})$ is a subset of $T_{\mathbb{R}}(f) \setminus S_r(f, \mathbb{R})$ and $S_{r+1}(f, \mathbb{R})$. We consider the only if part. Suppose that $\mathscr{T}_{r+1}(f, \mathbb{R}) = S_{r+1}(f) \setminus S_r(f)$ has no nonempty Euclidean open set. Then, $\mathrm{cl}\, S_{r+1}(f) = \mathrm{cl}\, S_r(f)$. Repeating Lemma 7.2, we have

$$\mathrm{cl}\, S_r(f) = \mathrm{cl}\, S_{r+1}(f) = \cdots = \mathrm{cl}\, S_{\mathrm{max.rank}(f)}(f) = T_{\mathbb{R}}(f),$$

and then, $T_{\mathbb{R}}(f) \setminus S_r(f, \mathbb{R})$ has no nonempty open set. Therefore, if $T_{\mathbb{R}}(f) \setminus S_r(f, \mathbb{R})$ includes a nonempty Euclidean open set, then $S_{r+1}(f) \setminus S_r(f)$ also includes a nonempty Euclidean open set.

In particular, for $r \in \mathrm{trank}_{\mathbb{R}}(f)$, if $T_{\mathbb{R}}(f) \setminus S_r(f, \mathbb{R})$ includes a Euclidean open set, then $r + 1 \in \mathrm{trank}_{\mathbb{R}}(f)$. We also have the following proposition.

Proposition 7.7 *For $r_1, r_2 \in \mathrm{trank}_{\mathbb{R}}(f)$ with $r_1 \leq r_2$, $s \in \mathrm{trank}_{\mathbb{R}}(f)$ for any s with $r_1 \leq s \leq r_2$.*

Proof Let s and t be the minimal and maximal typical rank of tensors with format f, respectively. By Lemma 7.2, we see that

$$\mathrm{cl}\, S_s(f) \subsetneqq \mathrm{cl}\, S_{s+1}(f) \subsetneqq \cdots \subsetneqq \mathrm{cl}\, S_t(f) = T_{\mathbb{R}}(f).$$

Thus, an arbitrary integer r with $s \leq r \leq t$ is a typical rank of $T_{\mathbb{R}}(f)$ by Proposition 7.6.

Following Strassen (1983) and Lickteig (1985), Bürgisser et al. (1997) obtained the asymptotic growth of the function grank and determined its value for some special formats. It is not easy to see whether the Jacobian has full column rank. The problem amounts to determining the dimension of higher secant varieties to Segre varieties. They achieved this by computing the dimension of the tangent space to these varieties, for which some machinery was developed.

Let $f = (f_1, f_2, f_3)$. For $t = u_1 \otimes u_2 \otimes u_3 \in S_1(f, \mathbb{C})$, we denote by $T_{\mathbb{C}}(f) \diamond_1 t$, $T_{\mathbb{C}}(f) \diamond_2 t$, $T_{\mathbb{C}}(f) \diamond_3 t$ the subspaces $\mathbb{C}^{f_1} \otimes u_2 \otimes u_3$, $u_1 \otimes \mathbb{C}^{f_2} \otimes u_3$, $u_1 \otimes u_2 \otimes \mathbb{C}^{f_3}$ of $T_{\mathbb{C}}(f)$, respectively. The sum of these three subspaces is the tangent space of $S_1(f, \mathbb{C})$ at t. A 4-tuple $s := (s_0; s_1, s_2, s_3) \in \mathbb{N}^4$ is called a *configuration*. If $(t; x, y, z) \in S_{|s|}(f, \mathbb{C}) := S_{s_0}(f, \mathbb{C}) \times S_{s_1}(f, \mathbb{C}) \times S_{s_2}(f, \mathbb{C}) \times S_{s_3}(f, \mathbb{C})$, we denote by $\Sigma_f(t; x, y, z)$ the following subspace of $T_{\mathbb{C}}(f)$:

$$\Sigma_f(t; x, y, z) := \sum_{k=1}^{s_0} (T_{\mathbb{C}}(f) \diamond_1 t_k + T_{\mathbb{C}}(f) \diamond_2 t_k + T_{\mathbb{C}}(f) \diamond_3 t_k)$$
$$+ \sum_{\alpha=1}^{s_1} T_{\mathbb{C}}(f) \diamond_1 x_\alpha + \sum_{\beta=1}^{s_2} T_{\mathbb{C}}(f) \diamond_2 y_\beta + \sum_{\gamma=1}^{s_3} T_{\mathbb{C}}(f) \diamond_3 z_\gamma,$$

where $t = \sum_{k=1}^{s_0} t_k \in S_{s_0}(f, \mathbb{C})$, $x = \sum_{\alpha=1}^{s_1} x_\alpha \in S_{s_1}(f, \mathbb{C})$, $y = \sum_{\beta=1}^{s_2} y_\beta \in S_{s_2}(f, \mathbb{C})$, and $z = \sum_{\gamma=1}^{s_3} z_\gamma \in S_{s_3}(f, \mathbb{C})$. The map $S_{|s|}(f, \mathbb{C}) \to \mathbb{N}$, $(t; x, y, z) \mapsto \dim \Sigma_f(t; x, y, z)$ is Zariski lower semi-continuous, i.e., the sets $\{(t; x, y, z) \mid \dim \Sigma_f(t; x, y, z) < r\}$ are Zariski open for all $r \in \mathbb{N}$. We denote the maximum value of the above map by $d(s, f)$ and call it the dimension of the configuration s in the format f. Note that, by semi-continuity, $d(s, f)$ is also the generic value of the above map. We easily see the following dimension estimation:

$$d(s, f) \leq \min\{s_0(f_1 + f_2 + f_3 - 2) + s_1 f_1 + s_2 f_2 + s_3 f_3, f_1 f_2 f_3\}.$$

Definition 7.1 A configuration s is said to *fill* a format f, $s \succ f$, if and only if $d(s, f) = f_1 f_2 f_3$. The configuration s is said to *exactly fill* f, $s \asymp f$, if and only if $d(s, f) = s_0(f_1 + f_2 + f_3 - 2) + s_1 f_1 + s_2 f_2 + s_3 f_3 = f_1 f_2 f_3$.

Lemma 7.3 (Bürgisser et al. 1997, (20.13) Lemma)

(1) *The relations $\succ$ and $\asymp$ are invariant under simultaneous permutation of the components of f and the last three components of s.*
(2) *if $S \geq s$, $f \geq F$ component-wise, then $s \succ f$ implies that $S \succ F$.*
(3) *$(r; 0, 0, 0) \succ f$ implies that $\mathrm{grank}(f) \leq r$.*
(4) *$(r; 0, 0, 0) \asymp f$ implies that f is perfect and $\mathrm{grank}(f) = r$.*

Lemma 7.4 (Bürgisser et al. 1997, (20.15) Lemma) *For all $a, b, c, d \in \mathbb{N} \cup \{0\}$, the following relations hold:*

(1) *$(1; 0, 0, 0) \asymp (1, 1, a)$, $(0; 0, 0, 1) \asymp (1, 1, a)$ if $a > 0$,*
(2) *$(0; bc, 0, 0) \asymp (a, b, c)$ if $abc > 0$,*
(3) *$(0; bc, ad, 0) \asymp (a, b, c + d)$ if $ab(c + d) > 0$,*
(4) *$(1; ab, 0, 0) \asymp (1, a + 1, b + 1)$,*
(5) *$(a; b, 0, 0) \asymp (a, 2, a + b)$ if $a > 0$,*
(6) *$(2a; 0, ab, 0) \asymp (2a + b, 2, 2a)$ if $a > 0$,*
(7) *$(2a; 0, 2ab + 2ac + 4bc, 0) \asymp (2a + 2b, 2, 2a + 2c)$ if $(a + b)(a + c) > 0$,*
(8) *$(2ad; 0, 0, 2a(b+c-d+1)+4bc) \asymp (2a+2b, 2a+2c, 2d)$ if $(a+b)(a+c)d > 0$ and $a(b + c - d + 1) + 2bc \geq 0$,*
(9) *$(2ad; 0, 0, 0) \succ (2a + 2b, 2a + 2c, 2d)$ if $(a + b)(a + c)d > 0$ and $a(b + c - d + 1) + 2bc \leq 0$.*

Theorem 7.6 *If $1 \leq f_1 \leq f_2 \leq f_3$, then*

$$\mathrm{grank}(2f_1, 2f_2, 2f_3) \leq 2f_3 \left\lceil \frac{2f_1 f_2}{f_1 + f_2 + f_3 - 1} \right\rceil.$$

In addition, if $2f_1 f_2/(f_1 + f_2 + f_3 - 1)$ is an integer, then $(2f_1, 2f_2, 2f_3)$ is perfect.

Proof Let $a = \lceil \frac{2f_1 f_2}{f_1 + f_2 + f_3 - 1} \rceil$, $b = f_1 - a$, $c = f_2 - a$, and $d = f_3$. We see that $a(b + c - d + 1) + 2bc = 2f_1 f_2 - a(f_1 + f_2 + f_3 - 1) \leq 0$. By Lemma 7.4 (9), $(2af_3; 0, 0, 0)$ is fill $(2f_1, 2f_2, 2f_3)$. Hence, by Lemma 7.3 (3), $\mathrm{grank}(2f_1, 2f_2, 2f_3) \leq 2f_3 a$.

Suppose that $2f_1 f_2/(f_1 + f_2 + f_3 - 2)$ is an integer. Then, $a(b+c-d+1)+2bc = 0$ and $(2ad; 0, 0, 0)$ is exactly fill $(2f_1, 2f_2, 2f_3)$ by Lemma 7.4 (8), and hence, the format $(2f_1, 2f_2, 2f_3)$ is perfect by Lemma 7.3 (4).

Corollary 7.1 (Strassen 1983, Bürgisser et al. 1997, (20.9) Theorem) *Suppose that f_1, f_2, f_3 are all even. If*

$$\frac{f_1 f_2 f_3}{(f_1 + f_2 + f_3 - 2) \max\{f_1, f_2, f_3\}}$$

is an integer, then (f_1, f_2, f_3) is perfect. For example, $(n, n, n + 2)$ is perfect if n is divisible by 6.

Theorem 7.7 (cf. Bürgisser et al. 1997, (20.9) Theorem) *Suppose that $f_1 \leq f_2 \leq f_3$.*

$$\lim_{f_1 \to \infty} \frac{\text{grank}(f_1, f_2, f_3)}{Q(f_1, f_2, f_3)} = 1.$$

Proof For $j = 1, 2, 3$, we take $\epsilon_j = 0, 1$ so that $f_j + \epsilon$ is even and put $g_j = 1 + \epsilon_j / f_j$. Propositions 1.1 and 7.4 (1) and Theorem 7.6 imply that

$$\text{grank}(f_1, f_2, f_3) \leq \min\{f_1 f_2, \text{grank}(f_1 g_1, f_2 g_2, f_3 g_3)\}$$

$$\leq \min\left\{f_1 f_2, \ f_3 g_3 \left\lceil \frac{f_1 g_1 f_2 g_2}{f_1 g_1 + f_2 g_2 + f_3 g_3 - 2} \right\rceil\right\}$$

$$\leq \min\left\{f_1 f_2, \ f_3 g_3 + \frac{f_1 g_1 f_2 g_2 f_3 g_3}{f_1 g_1 + f_2 g_2 + f_3 g_3 - 2}\right\}$$

Then, we see that

$$1 \leq \frac{\text{grank}(f_1, f_2, f_3)}{Q(f_1, f_2, f_3)} \leq \frac{(f_1 + f_2 + f_3 - 2)\text{grank}(f_1, f_2, f_3)}{f_1 f_2 f_3}$$

$$\leq \min\{1 + \frac{f_1 + f_2 - 2}{f_3}, \alpha\},$$

where

$$\alpha := \frac{g_3(f_1 + f_2 + f_3 - 2)}{f_1 f_2} + \frac{g_1 g_2 g_3 (f_1 + f_2 + f_3 - 2)}{f_1 g_1 + f_2 g_2 + f_3 g_3 - 2}.$$

If $f_3(f_1 f_2)^{-1}$ goes to 0, then α goes to 1 as f_1 goes to ∞. Otherwise, there exist constants $c, d > 0$ such that $f_3(f_1 f_2)^{-1} > c$ for any $f_1 > d$. Then, $(f_1 + f_2 - 2)f_3^{-1} < (cf_1)^{-1} + (cf_2)^{-1}$ if $f_1 > d$ and $1 + (f_1 + f_2 - 2)f_3^{-1}$ goes to 1. Therefore, $\text{grank}(f_1, f_2, f_3)Q(f_1, f_2, f_3)^{-1}$ goes to 1.

At the end of this section, we will show the special case of the following theorem by computing the rank of the Jacobian matrix.

Theorem 7.8 (Catalisano et al. 2008, Theorem 2.4) *Suppose that $2 \leq f_1 \leq \ldots \leq f_{n+1}$. Let $q = f_1 \cdots f_n - (f_1 + \cdots + f_n) + n$. If $f_{n+1} = q$, then f is perfect. If $q \leq f_{n+1}$, then $\text{grank}(f) = \min\{f_1 \cdots f_n, f_{n+1}\}$.*

We show this for $n = 2$. Let $2 \leq f_1 \leq f_2 \leq f_3$ and put $q = f_1 f_2 - f_1 - f_2 + 2$. For $0 < x \leq f_1 f_2$, the inequality $x - 1 < f_1 f_2 x / (f_1 + f_2 + q - 2) \leq x$ if and only if $(m - 1)(n - 1) + 1 \leq x \leq mn$. Then, for $q \leq f_3 \leq f_1 f_2$, $\text{grank}(f_1, f_2, f_3) \geq Q(f_1, f_2, f_3)$. For $f_3 = q$, then $f_1 f_2 f_3 / (f_1 + f_2 + f_3 - 2)$ is an integer. It suffices to show that $d_{Q(f_1, f_2, f_3)} = f_1 f_2 f_3$.

Suppose that $q \leq f_3 \leq f_1 f_2$. Let S_1 be a subset of

$$S = \{(k_1, k_2) \mid k_j \in \mathbb{N}, \ 1 \leq k_j \leq f_j, \ j = 1, 2\}$$

with cardinality f_3, which includes

$$S_1 = \{(k_1, k_2) \mid k_j \in \mathbb{N}, \; 1 \le k_j < f_j, \; j = 1, 2\} \cup \{(f_1, f_2)\},$$

and let $\iota: S_1 \to \{1, 2, \ldots, f_3\}$ be a bijection. We define maps $u_1, u_2: S_1 \to \mathbb{Z}$ by

$$u_h(x_1, x_2) = \begin{cases} 0 & \text{if } x_h = f_h, \\ 1 & \text{if } x_h < f_h \text{ and } x_{h'} = f_{h'} \text{ where } \{h, h'\} = \{1, 2\}, \\ \iota(x_1, x_2) + 1 & \text{otherwise} \end{cases}$$

for $h = 1, 2$. Let $e_j^{(h)}$ denote the jth row vector of the identity $f_h \times f_h$ matrix. We put $a_k^{(h)} \in \mathbb{R}^{f_h}, h = 1, 2, 3$, as

$$a_{\iota(k_1,k_2)}^{(h)} = e_{k_h}^{(h)} + u_h(k_1, k_2) e_{f_h}^{(h)}, \quad h = 1, 2$$

$$a_{\iota(k_1,k_2)}^{(3)} = e_{\iota(k_1,k_2)}^{(3)}$$

for all $(k_1, k_2) \in S_1$. Put

$$z = \left(a_1^{(1)}, a_1^{(2)}, a_1^{(3)}, a_2^{(1)}, a_2^{(2)}, a_2^{(3)}, \ldots, a_{f_3}^{(1)}, a_{f_3}^{(2)}, a_{f_3}^{(3)} \right) \in \mathbb{R}^{(f_1 + f_2 + f_3)f_3}.$$

It suffices to show that $J_{\Phi_{f_3}}(z) x^T = 0$ implies that $x = 0$.

Let $k = (k_1, k_2)$, $i = (i_1, i_2)$, and $I = \{(i_1, i_2) \mid 1 \le i_1 \le f_1, 1 \le i_2 \le f_2\}$. The equation $J_{\Phi_{f_3}}(z) x^T = 0$ for $x = (x(i_1, i_2, \iota(k)))_{i \in I, k \in S_1}$ is equivalent to the following equations

$$x(i_1, k_2, \iota(k)) + u_2(k) x(i_1, f_2, \iota(k)) = 0, \tag{7.2.1}$$

$$x(k_1, i_2, \iota(k)) + u_1(k) x(f_1, i_2, \iota(k)) = 0, \tag{7.2.2}$$

$$x(k_1, k_2, i_3) + u_1(k) x(f_1, k_2, i_3) + u_2(k) x(k_1, f_2, i_3)$$
$$+ u_1(k) u_2(k) x(f_1, f_2, i_3) = 0, \tag{7.2.3}$$

for $(i_1, i_2) \in I$, $1 \le i_3 \le f_3$, and $k \in S_1$. Since ι is bijective, by changing the symbols, Eq. (7.2.3) implies that

$$x(j_1, j_2, \iota(k)) + u_1(j) x(f_1, j_2, \iota(k)) + u_2(j) x(j_1, f_2, \iota(k))$$
$$+ u_1(j) u_2(j) x(f_1, f_2, \iota(k)) = 0, \tag{7.2.4}$$

for $j = (j_1, j_2) \in S_1$ and $k \in S_1$. By substituting $j = (f_1, f_2)$ in (7.2.4), we see that

$$x(f_1, f_2, \iota(k)) = 0 \tag{7.2.5}$$

for any $k \in S_1$. Let j_1 and j_2 be arbitrary integers such that $1 \leq j_1 < f_1$ and $1 \leq j_2 < f_2$. By Eqs. (7.2.5), (7.2.1), (7.2.2) and (7.2.4) imply that

$$x(j_1, k_2, \iota(k)) + u_2(k)x(j_1, f_2, \iota(k)) = 0, \quad (7.2.6)$$
$$x(f_1, k_2, \iota(k)) = 0, \quad (7.2.7)$$
$$x(k_1, j_2, \iota(k)) + u_1(k)x(f_1, j_2, \iota(k)) = 0, \quad (7.2.8)$$
$$x(k_1, f_2, \iota(k)) = 0, \quad (7.2.9)$$
$$x(j_1, j_2, \iota(k)) + u_1(j)x(f_1, j_2, \iota(k)) + u_2(j)x(j_1, f_2, \iota(k)) = 0. \quad (7.2.10)$$

By substituting $j_2 = k_2$, Eqs. (7.2.10) and (7.2.7) imply that

$$x(j_1, k_2, \iota(k)) + u_2(j_1, k_2)x(j_1, f_2, \iota(k)) = 0,$$

and in addition, by (7.2.8), we see that

$$(u_2(k_1, k_2) - u_2(j_1, k_2))x(j_1, f_2, \iota(k)) = 0.$$

If $j_1 \neq k_1$, then $x(j_1, f_2, \iota(k)) = 0$ since $u_2(k_1, k_2) \neq u_2(j_1, k_2)$ by definition. Thus, $x(j_1, f_2, \iota(k)) = 0$ for arbitrary j_1 with $1 \leq j_1 < f_1$ by (7.2.9). Similarly, by substituting $j_1 = k_1$, we have $x(f_1, j_2, \iota(k)) = 0$. Therefore, $x(j_1, j_2, \iota(k)) = 0$ by (7.2.10).

Consequently, we get $x = 0$, and thus, $J_{\Phi_{f_3}}(z)$ has full column rank, which implies that $d_{Q(f_1, f_2, f_3)} = f_1 f_2 f_3$.

Chapter 8
$2 \times 2 \times \cdots \times 2$ Tensors

In this chapter, we consider an upper bound of the rank of an n-tensor with format $(2, 2, \ldots, 2)$ over the complex and real number fields.

8.1   Introduction

Throughout this chapter, let $F_n = (2, 2, \ldots, 2)$.

Let $T = (A; B) = (t_{ijk})$ be a 3-tensor of $T_{\mathbb{K}}(F_3)$ and put $A = (a_1, a_2)$ and $B = (b_1, b_2)$. Cayley's hyperdeterminant $\Delta(T)$ of T is defined by

$$t_{111}^2 t_{222}^2 + t_{112}^2 t_{221}^2 + t_{121}^2 t_{212}^2 + t_{211}^2 t_{122}^2 - 2t_{111}t_{112}t_{221}t_{222} - 2t_{111}t_{121}t_{212}t_{222}$$
$$- 2t_{111}t_{122}t_{211}t_{222} - 2t_{112}t_{121}t_{212}t_{221} - 2t_{112}t_{122}t_{221}t_{211} - 2t_{121}t_{122}t_{212}t_{211}$$
$$+ 4t_{111}t_{122}t_{212}t_{221} + 4t_{112}t_{121}t_{211}t_{222}.$$

It is also described as

$$\Delta(T) = (\det(a_1, b_2) - \det(a_2, b_1))^2 - 4 \det(a_1, a_2) \det(b_1, b_2)$$

and $\Delta(T)$ is the discriminant of the quadratic polynomial

$$\det(A)x^2 - (\det(A + B) - \det(A) - \det(B))x + \det(B).$$

We have the following proposition straightforwardly.

Proposition 8.1 (de Silva and Lim 2008, Proposition 5.6)

$$\Delta((P, Q, R) \cdot (A; B)) = \Delta(A; B) \det(P)^2 \det(Q)^2 \det(R)^2$$

for any matrices P, Q, and R. In particular, $\Delta(A; B) = \Delta(B; A)$ and $\Delta(A + xB; yB) = y^2 \Delta(A; B)$ for any x and y.

© The Author(s) 2016
T. Sakata et al., *Algebraic and Computational Aspects of Real Tensor Ranks*,
JSS Research Series in Statistics, DOI 10.1007/978-4-431-55459-2_8

Cayley's hyperdeterminant is invariant under the action of $\mathrm{SL}(2, \mathbb{K})^{\times 3}$ and the sign of Cayley's hyperdeterminant is invariant under the action of $\mathrm{GL}(2, \mathbb{R})^{\times 3}$ if $\mathbb{K} = \mathbb{R}$.

The discriminant of the characteristic polynomial of $\det(A)A^{-1}B$ is equal to $\Delta(A; B)$. Thus, if $A^{-1}B$ has distinct eigenvalues, $\Delta(T)$ is positive if $\mathbb{K} = \mathbb{R}$ and nonzero in $\mathbb{K} = \mathbb{C}$. If A is nonsingular, $\mathrm{rank}(A; B) = 2$ if and only if $A^{-1}B$ is diagonalizable; thus, we have the following proposition.

Proposition 8.2 (de Silva and Lim 2008, Corollary 5.7, Propositions 5.9 and 5.10) *Let $T \in T_{\mathbb{R}}(F_3)$.*

(1) If $\Delta(T) > 0$, then $\mathrm{rank}(T) \leq 2$.
(2) If $\Delta(T) < 0$, then $\mathrm{rank}(T) = 3$.
(3) If $\mathrm{rank}(T) \leq 2$, then $\Delta(T) \geq 0$.

Theorem 8.1 (Sumi et al. 2014, Theorem 3) *Let $A = (a_1, a_2)$ and $B = (b_1, b_2)$ be 2×2 real (resp. complex) matrices and let $T = (A; B)$ be a tensor with format F_3. $\mathrm{rank}_{\mathbb{F}}(T) \leq 2$ if and only if*

(1) $\alpha A + \beta B = O$ for some $(\alpha, \beta) \neq (0, 0)$, or
(2) $\alpha(a_1, b_1) + \beta(a_2, b_2) = O$ for some $(\alpha, \beta) \neq (0, 0)$, or
(3) $\Delta(T) = 0$ and $\det(a_1, b_1) + \det(a_2, b_2) = 0$, or
(4) $\Delta(T)$ is positive (resp. nonzero).

We define a function $\Theta : T_{\mathbb{F}}(F_3) \to \mathbb{F}$ by

$$\Theta((a_1, a_2); (b_1, b_2)) = |a_1, b_1| + |a_2, b_2|.$$

We have the following corollary by Theorem 8.1.

Corollary 8.1 (Sumi et al. 2014, Corollary 1) *Let $T = ((a_1, a_2); (b_1, b_2))$ be a tensor with format F_3.*

(1) A complex tensor T has rank 3 if and only if

$$\dim \left\langle \begin{pmatrix} a_1 \\ a_2 \end{pmatrix}, \begin{pmatrix} b_1 \\ b_2 \end{pmatrix} \right\rangle = \dim \left\langle \begin{pmatrix} a_1 \\ b_1 \end{pmatrix}, \begin{pmatrix} a_2 \\ b_2 \end{pmatrix} \right\rangle = 2,$$

$\Delta(T) = 0$ and $\Theta(T) \neq 0$.
(2) A real tensor T has rank 3 if and only if $\Delta(T) < 0$, or

$$\dim \left\langle \begin{pmatrix} a_1 \\ a_2 \end{pmatrix}, \begin{pmatrix} b_1 \\ b_2 \end{pmatrix} \right\rangle = \dim \left\langle \begin{pmatrix} a_1 \\ b_1 \end{pmatrix}, \begin{pmatrix} a_2 \\ b_2 \end{pmatrix} \right\rangle = 2, \ \Delta(T) = 0 \ and \ \Theta(T) \neq 0.$$

Proposition 8.3 (Coolsaet 2013, Lemma 1) *Let $\mathbb{K}$ be a field and $A, B \in T_{\mathbb{K}}(2, 2)$. Then, $(A; B)$ is absolutely nonsingular if and only if $\det(A) \neq 0$ and the quadratic equation*

$$\det(A)x^2 - (\det(A + B) - \det(A) - \det(B))x + \det(B) = 0$$

has no solutions for x in $\mathbb{K}$. Equivalently, $(A; B)$ is absolutely nonsingular if and only if A is nonsingular and the eigenvalues of BA^{-1} do not belong to $\mathbb{K}$.

A tensor T of format F_3 over a field with characteristic $\neq 2$ is absolutely nonsingular if and only if $\Delta(T)$ is not a square in the field (see Coolsaet 2013).

8.2 Upper Bound of the Maximal Rank

Complex tensors with format F_n are an important target in quantum information theory (cf. see Verstraete et al. 2002).

A lower bound of the maximal rank of n-tensors over $\mathbb{F}$ with format F_n is

$$Q(F_n) = \left\lceil \frac{2^n}{2n - n + 1} \right\rceil = \left\lceil \frac{2^n}{n + 1} \right\rceil$$

(cf. Brylinski 2002, Proposition 1.2) and a canonical upper bound is 2^n. An upper bound using the maximal rank of tensors with format F_4 over $\mathbb{F}$ is known. The maximal rank of complex 4-tensors with format F_4 is just 4 (Brylinski 2002). The maximal rank of real 4-tensors with format F_4 is less than or equal to 5 (Kong and Jiang 2013). Nonsingular 4-tensors with format F_4 have been classified (Coolsaet 2013).

For a tensor $T = (t_{ijkl}) \in \mathbb{F}^{F_4}$, we write

$$T = ((T_{11}; T_{12}); (T_{21}; T_{22})) = \frac{T_{11} \mid T_{12}}{T_{21} \mid T_{22}} = \frac{\begin{array}{cc|cc} t_{1111} & t_{1211} & t_{1112} & t_{1212} \\ t_{2111} & t_{2211} & t_{2112} & t_{2212} \\ \hline t_{1121} & t_{1221} & t_{1122} & t_{1222} \\ t_{2121} & t_{2221} & t_{2122} & t_{2222} \end{array}}{}.$$

The action is given by $A_{ij} = PT_{ij}$ if $k = 1$, $A_{ij} = T_{ij}P$ if $k = 2$, $(A_{1j}; A_{2j}) = (T_{1j}; T_{2j}) \times_3 P$ if $k = 3$, and $(A_{i1}; A_{i2}) = (T_{i1}; T_{i2}) \times_3 P$ if $k = 4$, for $i, j = 1, 2$, where $((A_{11}; A_{12}); (A_{21}; A_{22})) = T \times_k P$.

Theorem 8.2 (Brylinski 2002, Theorem 1.1) *Any complex tensor with format F_4 has rank less than or equal to 4.*

Proof Let $T = ((T_{11}; T_{12}); (T_{21}; T_{22})) \in T_{\mathbb{C}}(F_4)$. If $\mathrm{rank}_{\mathbb{C}}(T_{11}; T_{12}) \leq 1$, then

$$\mathrm{rank}_{\mathbb{C}}(T) \leq \mathrm{rank}_{\mathbb{C}}(T_{11}; T_{12}) + \max.\mathrm{rank}_{\mathbb{C}}(F_3) \leq 4.$$

If T is $\mathrm{GL}(2, \mathbb{C})^{\times 4}$-equivalent to $T' = ((T'_{11}; T'_{12}); (T'_{21}; T'_{22}))$ with $\mathrm{rank}_{\mathbb{C}}(T'_{11}; T'_{12}) \leq 1$, then $\mathrm{rank}_{\mathbb{C}}(T) = \mathrm{rank}_{\mathbb{C}}(T') \leq 4$. Suppose that $\mathrm{rank}_{\mathbb{C}}(T'_{11}; T'_{12}) \geq 2$ for any tensor $T' = ((T'_{11}; T'_{12}); (T'_{21}; T'_{22}))$ that is $\mathrm{GL}(2, \mathbb{C})^{\times 4}$-equivalent to T.

First, suppose that $\mathrm{rank}_{\mathbb{C}}(T_{11}; T_{12}) = 2$. Then, $(T_{11}; T_{12})$ is $\mathrm{GL}(2, \mathbb{C})^{\times 3}$-equivalent

to $\left(\begin{pmatrix} 1 & 0 \\ 0 & 0 \end{pmatrix}; \begin{pmatrix} 0 & 1 \\ 0 & 0 \end{pmatrix} \right)$, $\left(\begin{pmatrix} 1 & 0 \\ 0 & 0 \end{pmatrix}; \begin{pmatrix} 0 & 0 \\ 1 & 0 \end{pmatrix} \right)$, $\left(\begin{pmatrix} 1 & 0 \\ 0 & 0 \end{pmatrix}; \begin{pmatrix} 0 & 0 \\ 0 & 1 \end{pmatrix} \right)$, or $\left(\begin{pmatrix} 1 & 0 \\ 0 & 1 \end{pmatrix}; O \right)$. We

consider the rank in each case. The first case is where $(T_{11}; T_{12})$ is $\mathrm{GL}(2, \mathbb{C})^{\times 3}$-

equivalent to $(C_1; C_2) := \left(\begin{pmatrix} 1 & 0 \\ 0 & 0 \end{pmatrix}; \begin{pmatrix} 0 & 1 \\ 0 & 0 \end{pmatrix} \right)$. The tensor T is $\mathrm{GL}(2, \mathbb{C})^{\times 4}$-equivalent

to $T' = ((T'_{11}; T'_{12}); (T'_{21}; T'_{22}))$ with $(T'_{11}; T'_{12}) = (C_1; C_2)$. If $\mathrm{rank}_{\mathbb{C}}(T'_{21} - x_0 C_1; T'_{22} - y_0 C_2) \leq 2$, then we see that

$$\mathrm{rank}_{\mathbb{C}}(T) \leq \mathrm{rank}_{\mathbb{C}}(T'_{21} + x_0 C_1; T'_{22} + y_0 C_2) + \mathrm{rank}_{\mathbb{C}}((C_1; O); (x_0 C_1; O))$$
$$+ \mathrm{rank}_{\mathbb{C}}((O; C_2); (O; y_0 C_2))$$
$$\leq 2 + 1 + 1 = 4,$$

since

$$T' = \frac{C_1 \; \big| \; C_2}{T'_{21} \; \big| \; T'_{22}} = \frac{O \; \big| \; O}{T'_{21} - x_0 C_1 \; \big| \; T'_{22} - y_0 C_2} + \frac{C_1 \; \big| \; O}{x_0 C_1 \; \big| \; O} + \frac{O \; \big| \; C_2}{O \; \big| \; y_0 C_2}.$$

To see that there exist x_0 and y_0 such that $\mathrm{rank}_{\mathbb{C}}(T'_{21} - x_0 C_1; T'_{22} - y_0 C_2) \leq 2$, we compute $\Theta(T'_{21} - x C_1; T'_{22} - y C_2)$ and $\Delta(T'_{21} - x C_1; T'_{22} - y C_2)$. If $\Theta(T'_{21} - x C_1; T'_{22} - y C_2) \neq 0$ for any x, y, then $b_{12} = a_{22} = 0$. Thus if $b_{12} \neq 0$ or $a_{22} \neq 0$ then $\mathrm{rank}_{\mathbb{C}}(T'_{21} - x_0 C_1; T'_{22} - y_0 C_2) \leq 2$ for some x_0, y_0 by Corollary 8.1 (1). The second

case is where $(T_{11}; T_{12})$ is $\mathrm{GL}(2, \mathbb{C})^{\times 3}$-equivalent to $(C_1; C_3) := \left(\begin{pmatrix} 1 & 0 \\ 0 & 0 \end{pmatrix}; \begin{pmatrix} 0 & 0 \\ 1 & 0 \end{pmatrix} \right)$.

There exist x_0, y_0 such that $\Theta(T'_{21} - x_0 C_1; T'_{22} - y_0 C_3) = 0$. Then, $\mathrm{rank}_{\mathbb{C}}(T'_{21} - x_0 C_1; T'_{22} - x_0 C_3) \leq 2$. The third case is where $(T_{11}; T_{12})$ is $\mathrm{GL}(2, \mathbb{C})^{\times 3}$-equivalent

to $(C_1; C_4) := \left(\begin{pmatrix} 1 & 0 \\ 0 & 0 \end{pmatrix}; \begin{pmatrix} 0 & 0 \\ 0 & 1 \end{pmatrix} \right)$. There exists x_0 such that $\Delta(T'_{21} - x_0 C_1; T'_{22} - x_0 C_4) \neq 0$, since $\Delta(T'_{21} - x_0 C_1; T'_{22} - x_0 C_4) = x^4 + o(x^3)$. Then, $\mathrm{rank}_{\mathbb{C}}(T'_{21} - x_0 C_1; T'_{22} - x_0 C_4) \leq 2$ by Corollary 8.1 (1). The fourth case is where $(T_{11}; T_{12})$ is $\mathrm{GL}(2, \mathbb{C})^{\times 3}$-equivalent to $(E_2; O)$. If $\mathrm{rank}_{\mathbb{C}}(T'_{21} - x C; T'_{22}) = 3$ for any x, then by considering the highest term of $\Theta(T'_{21} - x C; T'_{22})$ and $\Delta(T'_{21} - x C; T'_{22})$ for $C = C_1, C_4$, we have $T'_{22} = 0$, which is a contradiction. Therefore, if $\mathrm{rank}_{\mathbb{C}}(T_{11}; T_{12}) = 2$, then we have seen $\mathrm{rank}_{\mathbb{C}}(T) \leq 4$.

Next, suppose that $\mathrm{rank}_{\mathbb{C}}(T_{11}; T_{12}) = 3$. Then, $(T_{11}; T_{12})$ is $\mathrm{GL}(2, \mathbb{C})^{\times 3}$-equivalent

to $(E_2; C_2)$, where $C_2 = \begin{pmatrix} 0 & 1 \\ 0 & 0 \end{pmatrix}$. The tensor T is $\mathrm{GL}(2, \mathbb{C})^{\times 4}$-equivalent to

$T'' = ((E_2; C_2); (T''_{21}; T''_{22}))$ for some T''_{21}, T''_{22}. Since $\Theta(T''_{21} + x E_2; T''_{22} + x C_2) = -x^2 + o(x)$, $\mathrm{rank}_{\mathbb{C}}(T''_{21} + x E_2; T''_{22} + x C_2) = 2$ by Corollary 8.1 (1). Thus, by the above argument, we see that $\mathrm{rank}_{\mathbb{C}}(T) \leq 4$.

Corollary 8.2 For $n \geq 4$, the maximal rank of n-tensors with format F_n over the complex number field is less than or equal to 2^{n-2}.

Proof Theorem 8.2 covers the case where $n = 4$. Suppose that $n > 4$. The maximal rank of complex tensors with format F_4 is equal to 4. By applying Proposition 1.1, we have

$$\text{max.rank}_{\mathbb{C}}(F_n) \leq \text{max.rank}_{\mathbb{C}}(F_4) \prod_{t=5}^{n} 2 = 4 \cdot 2^{n-4} = 2^{n-2}.$$

Lemma 8.1 *Let n be a positive integer and let A_j and B_j, $1 \leq j \leq n$ be 2×2 real matrices. There exists a rank-1 real matrix C such that $\text{rank}_{\mathbb{R}}(A_j; B_j + C) \leq 2$ for any $1 \leq j \leq n$.*

Proof Put $A_j = \begin{pmatrix} a_j & b_j \\ c_j & d_j \end{pmatrix}$ and $C = \begin{pmatrix} su & sv \\ tu & tv \end{pmatrix}$. Since

$$\Delta(A_j; C) = (s(ud_j - vc_j) - t(ub_j - va_j))^2,$$

there exists a rank-1 matrix C_0 such that $\Delta(A_j; C_0) > 0$ for any $j \in S_2$. Let $C = \gamma C_0$. Since $(A_j; B_j + C)$ is $\{E_2\}^{\times 2} \times \text{GL}(2, \mathbb{R})$-equivalent to $(A_j; \gamma^{-1} B_j + C_0)$, the continuity of Δ implies that for each j, there exists $h_j > 0$ such that $\Delta(A_j; B_j + C) > 0$ for any $\gamma \geq h_j$ by Proposition 8.2 (1). For $C = (\max_j h_j)C_0$, we have $\text{rank}(A_j; B_j + C) \leq 2$ by Proposition 8.2 (2).

Theorem 8.3 (Sumi et al. 2014, Theorem 10) *Let $n \geq 2$. The maximal rank of real n-tensors with format F_n is less than or equal to $2^{n-2} + 1$.*

Proof The assertion is true for $n = 2, 3$. Then, suppose that $n \geq 4$. Let $e_{i_1,\ldots,i_n}$, $i_1, \ldots, i_n = 1, 2$ be a standard basis of $(\mathbb{R}^2)^{\otimes n}$, i.e., $e_{i_1,\ldots,i_n}$ has 1 as the $(i_1, \ldots, i_n)$-element and 0 otherwise. Any tensor A of $(\mathbb{R}^2)^{\otimes n}$ is written as

$$\sum_{i_1,\ldots,i_n} a_{i_1,\ldots,i_n} e_{i_1,\ldots,i_n}.$$

This is described as

$$\sum_{i_4,\ldots,i_n} B(i_4, \ldots, i_n) \otimes e_{i_4,\ldots,i_n},$$

where $B(i_4, \ldots, i_n) = \sum_{i_1,i_2,i_3} a_{i_1,\ldots,i_n} e_{i_1,i_2,i_3}$ is a tensor with format F_3. By Lemma 8.1, there exists a rank-1 2×2 matrix C such that $B(i_4, \ldots, i_n) + (O; C)$ has rank less than or equal to 2 for any $i_4, \ldots, i_n$. We have

$$A = \sum_{i_4,\ldots,i_n} (B(i_4, \ldots, i_n) + (O; C)) \otimes e_{i_4,\ldots,i_n} - \sum_{i_4,\ldots,i_n} (O; C) \otimes e_{i_4,\ldots,i_n}$$

$$= \sum_{i_4,\ldots,i_n} (B(i_4, \ldots, i_n) + (O; C)) \otimes e_{i_4,\ldots,i_n} - C \otimes e_2 \otimes u \otimes \cdots \otimes u,$$

where $u = \begin{pmatrix} 1 \\ 1 \end{pmatrix}$ and $e_2 = \begin{pmatrix} 0 \\ 1 \end{pmatrix}$, and then,

$$\text{rank}(A) \leq \sum_{i_4,\dots,i_n} \text{rank}(B(i_4,\dots,i_n) + (O; C)) + 1 = 2^{n-2} + 1.$$

Proposition 8.4 $\text{max.rank}_{\mathbb{R}}(n, n, n, n) \leq \frac{(n-1)(2n^2+2n+3)}{3}$.
$\text{max.rank}_{\mathbb{C}}(n, n, n, n) \leq \frac{2n^3+n-6}{3}$.

Proof We apply Proposition 5.4 to $\text{max.rank}_{\mathbb{F}}(n, n, n, n)$ twice. Then, we have

$$\text{max.rank}_{\mathbb{F}}(n, n, n, n) \leq n^2 + (n-1)^2 + \text{max.rank}_{\mathbb{F}}(n-1, n-1, n-1, n-1).$$

Thus, recursively, we have

$$\text{max.rank}_{\mathbb{F}}(n, n, n, n) \leq n^2 + 2\sum_{i=3}^{n-1} i^2 + 2^2 + \text{max.rank}_{\mathbb{F}}(F_4)$$

$$= \frac{2n^3 + n}{3} - 6 + \text{max.rank}_{\mathbb{F}}(F_4).$$

8.3 Typical Ranks

In this section, we consider the typical rank with format $F_n = (2, \dots, 2)$. The generic rank of n-tensors with format F_n is known. We have an upper bound of the maximal rank using the generic rank. Note that in general, the difference of the maximal rank and the generic rank is unbounded, e.g., $\text{max.rank}_{\mathbb{R}}(2n, 2n, 2) - \text{grank}(2n, 2n, 2) = n$.

Proposition 8.5 $\text{grank}(F_3) = 2$ *and* $\text{grank}(F_4) = 4$.

Recall that $\dim X_s(f) = \max_x \text{rank} J_{\Phi_s(f)}(x)$. We confirm the following property by Mathematica (see Table 8.1).

Proposition 8.6 $\max_x \text{rank} J_{\Phi_1(F_4)}(x) = 5$ *and* $\max_x \text{rank} J_{\Phi_2(F_4)}(x) = 10$, *but*

$$\max_x \text{rank} J_{\Phi_3(F_4)}(x) = 14 < 15.$$

In general, we have the following.

Theorem 8.4 (Catalisano et al. 2011, Theorem 4.1) $\dim \Sigma_s(F_n) = \min\{2^n, s(n+1)\}$ *for all* $n \geq 3$, $s \geq 1$ *except for* $n = 4$, $s = 3$. $\dim \Sigma_3(F_4) = 14$.

Table 8.1 Program for giving $\max_{x} \operatorname{rank} J_{\Phi_s(f)}(x)$

```
Jacob4[m_,n_,p_,q_,k_]:=Block[{i,j,mx,my,mz,mw,mm,x,y,z},
  For[i = 1,i ≤ k,i++,mx[i] = Array[x[i],m];my[i] = Array[y[i],n];
    mz[i] = Array[z[i],p];mw[i] = Array[w[i],q]];
  mm = Sum[Flatten[KroneckerProduct[Flatten[KroneckerProduct[
    Flatten[Transpose[{mx[i]}].{my[i]}],mz[i]],1],mw[i]],1],{i,1,k}]
  Flatten[Union[Table[D[mm,x[i][j]],{i,1,k},{j,1,m}],
    Table[D[mm,y[i][j]],{i,1,k},{j,1,n}],
    Table[D[mm,z[i][j]],{i,1,k},{j,1,p}],
    Table[D[mm,w[i][j]],{i,1,k},{j,1,q}]],1]]
For[i = 1,i ≤ 4,i++,a[i] = MatrixRank[Jacob4[2,2,2,2,i]]];
Print["J1=",a[1],", J2=",a[2],", J3=",a[3],", J4=",a[4]]
```

J1=5, J2=10, J3=14, J4=16

By this theorem, the generic rank of complex n-tensors with format F_n is equal to

$$Q(F_n) = \left\lceil \frac{2^n}{n+1} \right\rceil.$$

Theorem 8.5 (Blekherman and Teitler 2014) *Let f be an arbitrary format. The maximal rank of $T_{\mathbb{R}}(f)$ is less than or equal to twice the generic rank of $T_{\mathbb{C}}(f)$.*

$$\max.\operatorname{rank}_{\mathbb{R}}(f) \leq 2\operatorname{grank}(f).$$

Proof There exists a nonempty Euclidean open subset U of $T_{\mathbb{R}}(f)$ consisting of tensors with rank $\operatorname{grank}(f)$. Let $A \in U$ and consider $U' = \{-A+B \mid B \in U\}$. Then, U' is an open neighborhood of the zero tensor. For any $Y \in U'$, $\operatorname{rank} Y \leq 2\operatorname{grank}(f)$ by Proposition 1.1. Let X be any tensor of $T_{\mathbb{R}}(f)$. There exist $\varepsilon > 0$ and $Y \in U'$ such that $X = \varepsilon Y$. Since the rank is invariant under scalar multiplication, we see that $\operatorname{rank} X \leq 2\operatorname{grank}(f)$.

Recall that $\max.\operatorname{rank}_{\mathbb{R}}(f) = 3n$, $\operatorname{grank}(f) = 2n$, and $\operatorname{trank}_{\mathbb{R}}(f) = \{2n, 2n+1\}$ for $f = (2n, 2n, 2)$. Since $\operatorname{grank}(n, 2n, 3) = 2n$, $\max.\operatorname{rank}_{\mathbb{R}}(n, 2n, 3) \leq 4n$ by Theorem 8.5, but $\max.\operatorname{rank}_{\mathbb{R}}(n, 2n, 3) \leq 3n$. For sufficiently large n, it is not easy to estimate an upper bound of the maximal rank of the set of n-tensors.

Theorem 8.6 (Blekherman and Teitler 2014) *The maximal rank of the set of all n-tensors with format F_n is less than or equal to*

$$2Q(F_n) = 2\left\lceil \frac{2^n}{n+1} \right\rceil.$$

For a nonnegative integer r, $\mathscr{T}_r(f)$ denotes the set of all real tensors with format f and rank r.

Proposition 8.7 *Let $r \geq 1$. Any rank-1 tensor of $T_{\mathbb{F}}(f)$ lies in* cl $\mathscr{T}_r(f)$.

Proof Let $f = (f_1, \ldots, f_n)$. Let A be a rank-1 tensor. There exists $g_0 \in \mathrm{GL}(f_1, \mathbb{F}) \times \cdots \times \mathrm{GL}(f_n, \mathbb{F})$ such that $a_{1,\ldots,1} = 1$ and $a_{i_1,\ldots,i_n} = 0$ if $(i_1, \ldots, i_n) \neq (1, \ldots, 1)$,

where $(a_{i_1,\ldots,i_n}) = g_0 \cdot A$. For an integer $n \geq 1$, let $g_n = \prod_{j=1}^n \mathrm{diag}(\overbrace{1, 1/n, \ldots, 1/n}^{f_j}) \in \mathrm{GL}(f_1, \mathbb{F}) \times \cdots \times \mathrm{GL}(f_n, \mathbb{F})$. We see that $g_n \cdot \mathscr{T}_r(f) = \mathscr{T}_r(f)$ for $n \geq 0$ and that $g_0^{-1} g_n \cdot \mathscr{T}_r(f)$ converges to $\{xA \mid x \in \mathbb{F}\}$ as n goes to ∞.

For a subset V of $T_{\mathbb{R}}(f)$ and $A \in T_{\mathbb{R}}(f)$, we define $V - A := \{X - A \mid X \in V\}$. Then, it is easy to see that $\mathrm{cl}(V - A) = \mathrm{cl}(V) - A$ and $\mathrm{int}(V - A) = \mathrm{int}(V) - A$.

Theorem 8.7 *Let $A \in \mathrm{int}(\mathrm{cl}(\mathscr{T}_{\mathrm{grank}(f)}(f)))$ be a tensor of $T_{\mathbb{R}}(f)$. Any typical rank of $T_{\mathbb{R}}(f)$ is less than or equal to* $\mathrm{rank}_{\mathbb{R}}(A) + \mathrm{grank}(f)$.

Proof Let m, g, and a be the maximal typical rank of $T_{\mathbb{R}}(f)$, the generic rank of $T_{\mathbb{C}}(f)$, and $\mathrm{rank}_{\mathbb{R}}(A)$, respectively. Since $T_{\mathbb{R}}(f) \setminus \cup_{i=1}^m \mathscr{T}_i(f)$ is the union of semi-algebraic sets of dimension less than $\dim T_{\mathbb{R}}(f)$, it suffices to show that the set of tensors with rank less than or equal to $a + g$ is a dense subset of $T_{\mathbb{R}}(f)$. Furthermore, since the rank is invariant under scalar multiplication, it suffices to show that there exists an open set V such that $O \in V$ and $\mathrm{cl}(V \cap S_{a+g}(f)) = \mathrm{cl}(V)$.

Let U be an open set such that $A \in U$ and $U \subset \mathrm{int}(\mathrm{cl}(\mathscr{T}_g(f)))$. Put $V = U - A$. Since $A \in U$, V contains O. There exists a semi-algebraic subset S of $T_{\mathbb{R}}(f)$ such that $\dim S < \dim T_{\mathbb{R}}(f)$ and $\mathrm{rank}_{\mathbb{R}}(X) = g$ for any $X \in U \setminus S$. We see that $\mathrm{cl}((U \setminus S) - A) = \mathrm{cl}(V)$, since $(U \setminus S) - A = (U - A) \setminus (S - A)$. For any tensor Y in $(U \setminus S) - A$, $Y = Z + A$ for some $Z \in U \setminus S$ and $\mathrm{rank}_{\mathbb{R}}(Y) \leq \mathrm{rank}_{\mathbb{R}}(Z) + \mathrm{rank}_{\mathbb{R}}(A) = g + a$. Thus, $(U \setminus S) - A \subset S_{a+g}(f)$. Therefore, $\mathrm{cl}(V \cap S_{a+g}(f)) = \mathrm{cl}(V)$. This completes the proof.

By this theorem, we immediately have the following corollary.

Corollary 8.3 *Let $s = \min\{\mathrm{rank}_{\mathbb{R}}(A) \mid A \in \mathrm{int}(\mathrm{cl}(\mathscr{T}_{\mathrm{grank}(f)}(f)))\}$. Any typical rank of $T_{\mathbb{R}}(f)$ is less than or equal to $s + \mathrm{grank}(f)$. In particular, $T_{\mathbb{R}}(f)$ has a unique typical rank if $O \in \mathrm{int}(\mathrm{cl}(\mathscr{T}_{\mathrm{grank}(f)}(f)))$.*

If s might be equal to 1 in the above corollary, then the set of typical ranks of $T_{\mathbb{R}}(f)$ is a subset of $\{\mathrm{grank}(f), \mathrm{grank}(f) + 1\}$.

Let V_1 (resp. V_2) be the set consisting of all $(A; B) \in T_{\mathbb{R}}(n, n, 2)$ such that A is nonsingular and $A^{-1}B$ has distinct real (resp. distinct imaginary) eigenvalues. Then, $V_1 \subset \mathscr{T}_n(n, n, 2)$, $V_2 \subset \mathscr{T}_{n+1}(n, n, 2)$, and $\mathrm{cl}(V_1 \cup V_2) = T_{\mathbb{R}}(n, n, 2)$. The boundaries ∂V_1 and ∂V_2 are contained in $\{(A; B) \mid \mathrm{Res}(\det(\lambda A - B), \frac{d}{d\lambda} \det(\lambda A - B)) = 0\}$.

Theorem 8.8 *Let $A \in \mathrm{cl}(\mathscr{T}_{\mathrm{grank}(f)}(f)) \setminus \mathrm{int}(\mathrm{cl}(\mathscr{T}_{\mathrm{grank}(f)}(f)))$ be a nonzero tensor of $T_{\mathbb{R}}(f)$. Suppose that there exist $\varepsilon > 0$ and $X \in \mathrm{int}(\mathrm{cl}(\mathscr{T}_{\mathrm{grank}(f)}(f)))$ such that $B_\varepsilon(X) := \{Y \in T_{\mathbb{R}}(f) \mid ||Y - X|| < \varepsilon\} \subset \mathrm{int}(\mathrm{cl}(\mathscr{T}_{\mathrm{grank}(f)}(f)))$ and $A \in \mathrm{cl}(B_\varepsilon(X))$. Then, any typical rank of $T_{\mathbb{R}}(f)$ is less than or equal to $\mathrm{grank}(f) + \mathrm{rank}_{\mathbb{R}}(A)$.*

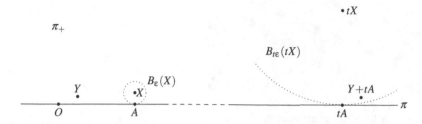

Fig. 8.1 The case where $Y \in \pi_+$

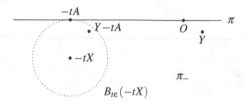

Fig. 8.2 The case where $Y \in \pi_-$

Proof Let $V = \text{cl}(\cup_{j=\text{grank}(f)}^{\text{grank}(f)+\text{rank}(A)} \mathcal{T}_j(f))$. It suffices to show that $V = T_{\mathbb{R}}(f)$. Note that $V = \text{cl}(S_{\text{grank}(f)+\text{rank}(A)}(f))$. Let π be a hyperplane that contains two points O and A, and is perpendicular to the line joining A and X, and let π_+ be the open half-space containing X separated by π in the space $T_{\mathbb{R}}(f)$.

Let $Y \in \pi_+$. There exists sufficiently large $t > 0$ such that $\varepsilon = ||X - A|| > ||X - A - t^{-1}Y)||$, i.e., $t\varepsilon = ||tX - tA|| > ||tX - (Y + tA)||$ and there exists an open neighborhood of $Y + tA$ that is a subset of $B_{t\varepsilon}(tX)$. We consider the set

$$U_+ = \{Y \in \pi_+ \mid Y + tA \in B_{t\varepsilon}(tX) \cap \mathcal{T}_{\text{grank}(f)}(f) \text{ for some } t > 0\}.$$

Then, since $B_{t\varepsilon}(tX) \subset \text{int}(\text{cl}(\mathcal{T}_{\text{grank}(f)}(f)))$, we see that $\pi_+ \subset \text{cl}(U_+)$ (see Fig. 8.1).

Next, let π_- be the half-space not containing X separated by π and let $Y \in \pi_-$. By a similar argument to the one above, we have $\pi_- \subset \text{cl}(U_-)$ (see Fig. 8.2), where

$$U_- = \{Y \in \pi_- \mid Y - tA \in B_{t\varepsilon}(-tX) \cap \mathcal{T}_{\text{grank}(f)}(f) \text{ for some } t > 0\}.$$

For $Z = Y + tA \in U_+ \cup U_-$, $\text{rank}_{\mathbb{R}}(Y) \le \text{rank}_{\mathbb{R}}(Z) + \text{rank}_{\mathbb{R}}(A) = \text{grank}(f) + \text{rank}_{\mathbb{R}}(A)$. This implies that $U_+ \cup U_- \subset S_{\text{grank}(f)+\text{rank}(A)}(f)$. Therefore,

$$T_{\mathbb{R}}(f) \setminus \pi = \pi_+ \cup \pi_- \subset \text{cl}(U_+ \cup U_-) \subset V,$$

and thus, $T_{\mathbb{R}}(f) = V$, since V is a closed set.

References

Adams, J. F. (1962). Vector fields on spheres. *Annals of Mathematics, 75*(2), 603–632.

Artin, E. (1927). Über die Zerlegung definiter Funktionen in Quadrate. *Abhandlungen aus dem Mathematischen Seminar der Universitat Hamburg, 5*(1), 100–115.

Atkinson, M. D., & Lloyd, S. (1980). Bounds on the ranks of some 3-tensors. *Linear Algebra and its Applications, 31*, 19–31.

Atkinson, M. D., & Lloyd, S. (1983). The ranks of $m \times n \times (mn - 2)$ tensors. *SIAM Journal on Computing, 12*(4), 611–615.

Atkinson, M. D., & Stephens, N. M. (1979). On the maximal multiplicity complexity of a family of bilinear forms. *Linear Algebra and its Applications, 27*, 1–8.

Bailey, R. A., & Rowley, C. A. (1993). Maximal rank of an element of a tensor product. *Linear Algebra and its Applications, 182*, 1–7.

Bi, S. (2008). *A criteiron for a generic $m \times n \times n$ to have rank n.* Unpublished paper.

Blekherman, G., & Teitler, Z. (2015). On maximum, typical and generic ranks. *Annals of Mathematics, 362*(3), 1021–1031.

Bochnak, J., Coste, M., & Roy, M.-F. (1998). Real algebraic geometry. Volume 36 of Ergebnisse der Mathematik und ihrer Grenzgebiete (3). Berlin: Springer-Verlag. Translated from the 1987 French original, Revised by the authors.

Bosch, A. J. (1986). The factorization of a square matrix into two symmetric matrices. *The American Mathematical Monthly, 93*(6), 462–464.

Brockett, R. W., & Dobkin, D. (1978). On the optimal evaluation of a set of bilinear forms. *Linear Algebra and its Applications, 19*(3), 207–235.

Brylinski, J. L. (2002). Algebraic measures of entanglement. *Chapter I in mathematics of quantum computation* (pp. 3–23). Boca Raton: Chapman and Hall/CRC.

Bürgisser, P., Clausen, M., & Shokrollahi, M. A. (1997). Algebraic complexity theory, Volume 315 of Grundlehren der Mathematischen Wissenschaften. Berlin: Springer. With the collaboration of Thomas Lickteig.

Catalisano, M. V., Geramita, A. V., & Gimigliano, A. (2008). On the ideals of secant varieties to certain rational varieties. *Journal of Algebra, 319*(5), 1913–1931.

Catalisano, M. V., Geramita, A. V., & Gimigliano, A. (2011). Secant varieties of $\mathbb{P}^1 \times \cdots \times \mathbb{P}^1$ (n-times) are not defective for $n \geq 5$. *Journal of Algebraic Geometry, 20*(2), 295–327.

Chevalley, C. (1955–1956). Schémas normaux; morphismes; ensembles constructibles. *Séminaire Henri Cartan 8 exposé 7.*

Cichocki, A., Zdunek, R., Phan, A. H., & Amari, S.-I. (2009). *Nonnegative matrix and tensor factorizations: Applications to exploratory multi-way data analysis and blind source separation.* New York: Wiley.

© The Author(s) 2016   103
T. Sakata et al., *Algebraic and Computational Aspects of Real Tensor Ranks*,
JSS Research Series in Statistics, DOI 10.1007/978-4-431-55459-2

Comon, P., ten Berge, J. M. F., De Lathauwer, L., & Castaing, J. (2009). Generic and typical ranks of multi-way arrays. *Linear Algebra and its Applications, 430*(11–12), 2997–3007.

Coolsaet, K. (2013). On the classification of nonsingular $2 \times 2 \times 2 \times 2$ hypercubes. *Designs, Codes and Cryptography, 68*(1–3), 179–194.

Cox, D., Little, J., & O'Shea, D. (1992). Ideals, varieties, and algorithms. An introduction to computational algebraic geometry and commutative algebra. *Undergraduate texts in mathematics.* New York: Springer.

De Lathauwer, L. (2006). A link between the canonical decomposition in multilinear algebra and simultaneous matrix diagonalization. *SIAM Journal on Matrix Analysis and Applications, 28*(3), 642–666 (electronic).

De Lathauwer, L., De Moor, B., & Vandewalle, J. (2000). A multilinear singular value decomposition. *SIAM Journal on Matrix Analysis and Applications, 21*(4), 1253–1278 (electronic).

de Loera, J. A., & Santos, F. (1996). An effective version of Pólya's theorem on positive definite forms. *Journal of Pure and Applied Algebra, 108*(3), 231–240.

de Silva, V., & Lim, L.-H. (2008). Tensor rank and the ill-posedness of the best low-rank approximation problem. *SIAM Journal on Matrix Analysis and Applications, 30*(3), 1084–1127.

Friedland, S. (2012). On the generic and typical ranks of 3-tensors. *Linear Algebra and its Applications, 436*(3), 478–497.

Gantmacher, F. R. (1959). *The theory of matrices* (Vols. 1, 2). Translated by K. A. Hirsch. New York: Chelsea Publishing Co.

Geramita, A. V., & Seberry, J. (1979). *Orthogonal designs: Quadratic forms and hadamard matrices* (Vol. 45). Lecture notes in pure and applied mathematics. New York: Marcel Dekker Inc.

Harris, J. (1992). *Algebraic geometry. A first course.* Graduate texts in mathematics (Vol. 133). New York: Springer.

Harshman, R. A. (1970). *Foundations of the PARAFAC procedure: Models and conditions for an "explanatory" multimodal factor analysis.* UCLA working papers in phonetics (Vol. 16, pp. 1–84).

Heintz, J., & Sieveking, M. (1981). Absolute primality of polynomials is decidable in random polynomial time in the number of variables. In *Automata, languages and programming (Akko, 1981).* Lecture notes in computer science (Vol. 115, pp. 16–28). Berlin: Springer.

Hitchcock, F. L. (1927). The expression of a tensor or a polyadic as a sum of products. *Journal of Mathematical Physics, 6*(1), 164–189.

Howell, T. D. (1978). Global properties of tensor rank. *Linear Algebra and its Applications, 22,* 9–23.

Ja'Ja', J. (1979). Optimal evaluation of pairs of bilinear forms. *SIAM Journal on Computing, 8*(3), 443–462.

Kolda, T. G., & Bader, B. W. (2009). Tensor decompositions and applications. *SIAM Review, 51*(3), 455–500.

Kong, X., & Jiang, Y.-L. (2013). A note on the ranks of $2 \times 2 \times 2$ and $2 \times 2 \times 2 \times 2$ tensors. *Linear Multilinear Algebra, 61*(10), 1348–1362.

Kruskal, J. B. (1977). Three-way arrays: rank and uniqueness of trilinear decompositions, with application to arithmetic complexity and statistics. *Linear Algebra and its Applications, 18*(2), 95–138.

Landsberg, J. M. (2012). *Tensors: Geometry and applications.* Graduate studies in mathematics (Vol. 128). Providence: American Mathematical Society.

Lasserre, J. B. (2010). *Moments, positive polynomials and their applications.* Imperial college press optimization series (Vol. 1). London: Imperial College Press.

Lickteig, T. (1985). Typical tensorial rank. *Linear Algebra and its Applications, 69,* 95–120.

Lim, L.-H. (2014). Tensors and hypermatrices. In L. Hogben (Ed.) *Handbook of linear algebra* (2nd ed.). Discrete mathematics and its applications (Boca Raton). Boca Raton: CRC Press.

Marshall, M. (2008). *Positive polynomials and sums of squares.* Mathematical surveys and monographs (Vol. 146). Providence: American Mathematical Society

Matsumura, H. (1989). *Commutative ring theory* (2nd ed.). Cambridge Studies In Advanced Mathematics (Vol. 8). Cambridge: Cambridge University Press. Translated from the Japanese by M. Reid.

Miyazaki, M., Sumi, T., & Sakata T. (2009). Tensor rank determination problem. In *Proceedings CD International Conference Non Linear Theory and Its Applications* (pp. 391–394).

Mørup, M. (2011). *Applications of tensor (multiway array) factorizations and decompositions in data mining*. Wiley interdisciplinary reviews: Data mining and knowledge discovery (Vol. 1). New York: Wiley.

Mumford, D. (1976). *Algebraic geometry I: Complex projective varieties*. *Classics in mathematics*. Berlin: Springer.

Northcott, D. G. (1980). *Affine sets and affine groups*. London mathematical society lecture note series (Vol. 39). Cambridge: Cambridge University Press.

Northcott, D. G. (2008). *Multilinear algebra*. Cambridge: Cambridge University Press. (Reprint of the 1984 original).

Ottaviani, G. (2013). Introduction to the hyperdeterminant and to the rank of multidimensional matrices. *Commutative algebra* (pp. 609–638). New York: Springer.

Powers, V. (2011). Positive polynomials and sums of squares: Theory and practice. *Conference: Real Algebraic Geometry, Universite de Rennes* (Vol. 1).

Powers, V., & Reznick, B. (2001). A new bound for Pólya's theorem with applications to polynomials positive on polyhedra. *Journal of Pure and Applied Algebra, 164*(1–2), 221–229. (Effective methods in algebraic geometry (Bath, 2000)).

Powers, V., & Reznick, B. (2005). Polynomials positive on unbounded rectangles. *Positive polynomials in control*. Lecture notes in control and information sciences (Vol. 312, pp. 151–163). Berlin: Springer.

Prestel, A., & Delzell, C. N. (2001). *Positive polynomials: From Hilbert's 17th problem to real algebra*. Springer monographs in mathematics. Berlin: Springer.

Rangel, P. (2009). An introduction to some basic concepts on real algebraic geometry. Technical Reports.

Sakata, T., Maehara, K., Sasaki, T., Sumi, T., Miyazaki, M., Watanabe, Y., et al. (2012). Tests of inequivalence among absolutely nonsingular tensors through geometric invariants. *Universal Journal of Mathematics and Mathematical Sciences, 1*, 1–28.

Schmüdgen, K. (1991). The K-moment problem for compact semi-algebraic sets. *Mathematische Annalen, 289*(2), 203–206.

Schweighofer, M. (2002). An algorithmic approach to Schmüdgen's Positivstellensatz. *Journal of Pure and Applied Algebra, 166*(3), 307–319.

Shapiro, D. B. (2000). *Compositions of quadratic forms*. de Gruyter Expositions in Mathematics (Vol. 33). Berlin: Walter de Gruyter & Co.

Stengle, G. (1974). A nullstellensatz and a positivstellensatz in semialgebraic geometry. *Mathematische Annalen, 207*, 87–97.

Strassen, V. (1983). Rank and optimal computation of generic tensors. *Journal of Pure and Applied Algebra, 52*(53), 645–685.

Sumi, T., Miyazaki, M., & Sakata, T. (2009). Rank of 3-tensors with 2 slices and Kronecker canonical forms. *Journal of Pure and Applied Algebra, 431*(10), 1858–1868.

Sumi, T., Miyazaki, M., & Sakata, T. (2010). About the maximal rank of 3-tensors over the real and the complex number field. *Annals of the Institute of Statistical Mathematics, 62*(4), 807–822.

Sumi, T., Miyazaki, M., & Sakata, T. (2015a). Typical ranks of $m \times n \times (m - 1)n$ tensors with $3 \leq m \leq n$ over the real number field. *Linear Multilinear Algebra, 63*(5), 940–955.

Sumi, T., Miyazaki, M., & Sakata, T. (2015b). Typical ranks for 3-tensors, nonsingular bilinear maps and determinantal ideals. arXiv:1512.08452.

Sumi, T., Sakata, T., & Miyazaki, M. (2013). Typical ranks for $m \times n \times (m - 1)n$ tensors with $m \leq n$. *Linear Algebra and its Applications, 438*(2), 953–958.

Sumi, T., Sakata, T., & Miyazaki, M. (2014). Rank of n-tensors with size $2 \times \cdots \times 2$. *Far East Journal of Mathematical Sciences, 90*, 141–162.

ten Berge, J. M. F. (2000). The typical rank of tall three-way arrays. *Psychometrika*, *65*(4), 525–532.
ten Berge, J. M. F. (2011). Simplicity and typical rank results for three-way arrays. *Psychometrika*, *76*(1), 3–12.
ten Berge, J. M. F., & Kiers, H. A. L. (1999). Simplicity of core arrays in three-way principal component analysis and the typical rank of $p \times q \times 2$ arrays. *Linear Algebra and its Applications*, *294*(1–3), 169–179.
Verstraete, F., Dehaene, J., de Moor, B., & Verschelde, H. (2002). Four qubits can be entangled in nine different ways. *Physical Review A*, *65*(5), 052112. quant-ph/0109033v2.pdf.

Index

© The Author(s) 2016
T. Sakata et al., *Algebraic and Computational Aspects of Real Tensor Ranks*,
JSS Research Series in Statistics, DOI 10.1007/978-4-431-55459-2

Printed in the United States
B. Bookmasters

Printed in the United States
By Bookmasters